PRAISE FOR

Change Your Genes, Change Your Life

To Henning —
A votre santé —
Kenneth R. Pelletier

"An integrative medicine pioneer shares cutting-edge insights into the power of epigenetics to upgrade the genetic cards of life you have been dealt."

—Mehmet Oz, MD
Professor, New York Presbyterian Columbia;
Emmy Award-Winning Host, The Dr. Oz Show

"The future of healthcare will be preventable, personalized, predictable, and participatory. Read this book to find out how."

—Deepak Chopra, MD
Founder of the Chopra Center; Clinical Professor of Medicine, University of California School of Medicine at San Diego

"If you have ever worried that the genetic hand you have been dealt determines your destiny, worry no more. Dr. Pelletier's book redefines your genes from something you are stuck with to something you have profound influence over by the choices you make every day. If you want to be empowered to be the full expression of yourself, read this book."

—Mark Hyman, MD
New York Times best-selling author of *Food: What the Heck Should I Eat?*;
Director of the Cleveland Clinic Center for Functional Medicine

"In a health literate and innovative way, Dr. Pelletier unravels the mystery of your genes and the epigenetic inputs throughout life that continue to program your genes. Reading this book is an essential epigenetic input that will help you optimize your gene expression."

—Richard Carmona, MD, MPH, FACS
Seventeenth Surgeon General of the United States;
Distinguished Professor, University of Arizona

"This book revolutionizes the belief that we are inexorably controlled by our genes, and provides the same opportunity for redefining health as Dr. Pelletier's landmark *Mind as Healer, Mind as Slayer* did for the body-mind revolution in 1977."

—Jeffrey Bland, PhD
President, Personalized Lifestyle Medicine Institute

"Not only does this book provide a scientific basis for our ability to shape our health trajectories through belief and lifestyle choices, but it also outlines the emerging power of personalized medicine. It's a must-read for everybody interested in integrative medicine!"

—Emeran A. Mayer, MD, PhD
Executive Director, Oppenheimer Center for Neurobiology of Stress & Resilience;
Professor of Medicine at UCLA; author of *The Mind-Gut Connection*

"So you think health is all in your genes? Think again. It is in your epigenes! Dr. Pelletier once again show he is a master translator of this complex science into the simple and useable. Read it. Use it."

—Wayne Jonas, MD
Executive Director, Samueli Integrative Health Programs; Former Director,
Office of Alternative Medicine at the NIH; author of *How Healing Works*

"I found genetics to be the most complex subject in medical school. Now, fifty years later, the field is one hundred times more complicated. But leave it to Kenneth Pelletier to once again not only tackle the importance of this rapidly evolving field but, as with his other timely books, to make the practical application of these breakthroughs easy to understand for most readers. No one does it better."

—Steven E. Locke, MD
Associate Clinical Professor of Psychiatry at Harvard Medical School; Chief Medical Officer, iHope Network

"Dr. Pelletier continues in his visionary streak of anticipating and informing people about critical emerging trends in health promotion. In this newest book, he turns his sights on understanding the potential of lifestyle choices on changing the expression of the genes that shape our health. It is refreshing to see precision medicine focused on lifestyle change as an underappreciated complement to the development of pharmaceuticals. The book distills the complexities of genetics and metabolism into easy-to-understand concepts, metaphors, and examples that make these often arcane topics accessible and applicable to practical health improvement solutions."

—David S. Sobel, MD, MPH
Adjunct Lecturer, Stanford University School of Medicine; Former Director of Patient Education and Health Promotion, Kaiser Permanente Northern California; author of *The Mind & Body Health Handbook*

"Dr. Pelletier has been a trusted colleague for many years as well as an inspiration on the journey for health and wellness. In this book he has tackled the difficult subject of genomics and given us all hope that we still have control over our destinies as they relate to health. The introduction of any new technology requires the science to mature in order to provide a more complete understanding of its relevance and appropriate use. Dr. Pelletier has taken this information and broken it down into an interesting, informative read complete with advice on optimizing our own health. I am confident readers will walk away from this groundbreaking book reassured that lifestyle and environmental changes can make a big difference in each of our lives."

—K. Andrew Crighton, MD
Past Chairman of the Health Enhancement Research Organization

"The mapping of the human genome was truly one of the greatest scientific undertakings of the past century, detailing with incredible accuracy the blueprint of our species. It also paved the way for the field of epigenetics, which has shown that when it comes to our genes, *nurture* is inextricably linked with *nature.* In his new book, Dr. Pelletier, a true pioneer in mind-body medicine and integrative health research, makes a compelling case for why understanding our own unique genetic makeup can allow us to each make lifestyle and medical choices that can truly alter the trajectory of our lives. Scientifically based, informative, and thought-provoking, this book is for anyone interested in optimizing their health."

—Tieraona Low Dog, MD

Professor of Medicine, University of New Mexico School of Medicine;
Author of *National Geographic's Fortify Your Life*

"This extraordinary guidebook distills cutting-edge science and is beautifully written with practical tools for achieving optimal health. A must-read for consumers and clinicians."

—Woodson Merrell, MD

Assistant Professor of Medicine, Mt. Sinai School of Medicine;
Chairperson, Integrative Healthcare Symposium

"Healthy aging has been my passion for over 40 years. During that time I have always found the pioneering research and writings of my longtime friend Ken Pelletier to be invaluable. Now with *Change Your Genes*, he leads the way again into the realm of the emerging science of epigenetics with practical insights for all of us to attain optimal health and longevity. I highly recommend this book to anyone hoping to lead a healthy long life."

—Ken Dychtwald, PhD

Author of *A New Purpose: Redefining Money, Family, Work, Retirement, and Success*

"Dr. Pelletier's new book is practically giving away Ferraris—when it comes to sitting in the driver's seat of your own vehicle and steering an enjoyable journey to vibrant health. He has integrated decades of experience and research on how our genes truly respond to our lifestyle, including the role of stress, consciousness, and gut health. And, he offers us our own dashboard of markers, guidance systems, and practices to fulfill our potential as radiant, thriving beings."

—Foster Gamble
Co-creator, THRIVE Movie and Movement

CHANGE YOUR GENES, CHANGE YOUR LIFE

CREATING OPTIMAL HEALTH WITH THE NEW SCIENCE OF EPIGENETICS

CHANGE YOUR GENES CHANGE YOUR LIFE

DR. KENNETH R. PELLETIER

FOREWORD BY DR. ANDREW WEIL

ORIGIN PRESS

Publication date: October 2018

Origin Press
PO Box 151117
San Rafael, CA 94915
www.OriginPress.org

Cover and interior design by *the*BookDesigners.com

Library of Congress Card Catalog Number:
2018945176

ISBN: 978-1-57983-056-4

Publisher's Cataloging-in-Publication data:

Names: Pelletier, Kenneth R., author. | Weil, Andrew, foreword author.
Title: Change your genes , change your life : creating optimal health with the new science of epigenetics / Dr. Kenneth R. Pelletier ; foreword by Dr. Andrew Weil.

Description: Includes bibliographical references | San Rafael, CA: Origin Press, 2018.
Identifiers: ISBN 978-1-57983-056-4 (pbk.) | 978-1-57983-057-1 (ebook)

Subjects: LCSH Epigenetics--Health aspects. | Gene expression. | Cells--Morphology. | Genetic screening. | Public health. | Medical genetics--Popular works. | Preventive medicine. | Health. | BISAC MEDICAL / Genetics | SCIENCE / Life Sciences / Genetics & Genomics

Classification: LCC QH450 .P45 2018 | DDC 616/.042--dc23

First printing July 2018

Printed in the United States of America
10 9 8 7 6 5 4 3 2 1

DEDICATION

To my parents:
Roger N. Pelletier and Lucy B. Pelletier
Who gave me their gifts of compassion,
honesty, charity, hard work, love . . .
and life itself.

ACKNOWLEDGMENTS

There have been many individuals who have had a direct and indirect role in the writing of this book over several years. First of all, my deepest gratitude to my editor, Byron Belitsos, for his insights, perseverance, research skills, and writing ability to make this esoteric subject into a personal odyssey for the reader. Also a respectful thank-you to my agent, James Levine, for his patience and guidance in finding the right publisher for this manuscript. Of course I need to extend a heartfelt thank-you to my friend, Ms. Kathrin E. Nikolussi, for her unwavering support and encouragement to see my way through the long and winding road leading to this book.

Among the many colleagues who have influenced my thinking and made invaluable contributions to the research and knowledge in this book, I wish to thank Paula Nenn, Richard Carmona, David Sobel, Steve Locke, Rachel Naomi Remen, Eric Topol, Tieroana Low Dog, Mehmet Oz, Brian and Sue Berman, Deepak Chopra, Mark Hyman, Michael McGinnis, Regina Herzlinger, Jeff Bland, C. Norman Shealy, Joan Schleicher, Brent Bauer, Victoria Maizes, Joe Helms,

Emeran Mayer, Steve Schroeder, John Weeks, James Dalen, Bruce Lipton, Roger Morrison, Ken Dychtwald, Patrick Hanaway, Wayne Jonas, Jeff Davis, Larry and Barbara Dossey, Elizabeth Blackburn, Alain Enthoven, Mark Liponis, Michael O'Donnell, Don Berwick, Margaret Chesney, Barrie Cassileth, Mimi Guarneri, Tracy Gaudet, Michael Lerner, Cathy Baase, Jon Kabat-Zinn, Michael Schmidt, Andy Crighton, Robert Rountree, Dexter Shurney, Dave Thom, Ron Goetzel, Justine Greene, Seth Serxner, Tony Elite, Foster Gamble, Margaret Chesney, David Peters, Michael Finkelstein, Doug Metz, MJ Osmick, George and Jan DeVries, Jonathan Fielding, Woody Merrell, Marty Rossman, Jack Farquhar, Bill Haskell, John Sailer, Ben Kligler, Dean Ornish, Heather Tick, Jim Gordon, Arya Neilsen, Jim Fries, Denise Herzing, Jeri Ryan, Brent Bauer, Daniel Kraft, Robert Bonakdar, and of course Elizabeth A. Pelletier. Together these pioneering individuals have and are creating a true health care system for all of the people they influence and touch as well as for the entire planet.

CONTENTS

FOREWORD

By Dr. Andrew Weil

Over the past four decades, my friend Dr. Kenneth R. Pelletier has devoted his professional life to the study of what constitutes health and wellness. His entire body of work, including over a dozen books, has been a series of signposts pointing the way to a well-lived and healthful life. From diet and exercise to alternative medicine, and from acupuncture and meditation to the Fortune 500 corporate health programs that he was among the first to develop, Ken has always tried to show how our personal choices, when they are based in the latest research science, can lead to a healthier and hopefully longer life.

In concert with my own work over these decades, all of Ken's books and research and his many keynote talks given all over the world have encouraged us to consider not only the influence the mind has on health but also to explore our own consciousness for its own sake. He has urged us to ask ourselves who we are and what more is possible for all of us in terms of wellness, happiness, and longevity. He has always endeavored to show how our minds and bodies are an expression of a greater consciousness at work.

Epigenetics, the subject of this book, is but another marker buoy in the maritime channel that validates the importance of these previous considerations and perspectives, especially as they initially appeared in Ken's often-quoted classic *Mind As Healer, Mind As Slayer* (1977).

Ken's intent has always been to lead us to discover what *more* we can do to make healthier choices. Epigenetics now offers us the promise of far more specific health choices tailored to the special needs of each individual according to the unique characteristics of their genome. In Ken's hands, this revolutionary new science provides yet another luminous signpost pointing to the untapped potential of the human mind. It confirms once again what we have long suspected, that what we believe and how we think and behave has a literal impact on matter—that is, on the body's highly complex biochemical processes. While the right equation may not be "mind over matter," it is profoundly true, as Ken likes to say, that *our mind matters*! That's why in this book he asks this crucial question: "What does epigenesis tell us about our own untapped potentials and our self-limiting beliefs?" Answering it, and coming up with health-optimizing practices based on that answer, is a central purpose of this effort.

Epigenetics is now at the forefront of modern medicine, with new studies and findings pouring out almost daily. According to Dr. Pelletier, the real upshot of the epigenetic revolution for healthcare is that it opens the door to what futurists call *personalized medicine.* For perhaps the first time in an introductory trade book, he explains in layperson's language the genetic biomarkers that will become the standard reference for measuring which specific lifestyle changes are required to optimize your health. In the very near future, he explains, a state-of-the-art genetic and epigenetic profile—matched with other precise indicators such as assays of the gut microbiome—will guide our daily health practices. This is an exciting methodology for diagnosis and treatment that Ken has helped to pioneer. Ultimately, I believe it constitutes a promising new form of medicine that offers unprecedented hope for all of us focused on optimal, integrative health and longevity.

INTRODUCTION

In early 2001, I can vividly remember my excitement, and the enthusiasm of my medical colleagues, when the leaders of the Human Genome Project announced that they had achieved the "sequencing" of the complete human genome, thereby providing us with the first detailed map of the genetic code that resides in the nucleus of every cell.

Scientists and doctors are not known for expressing lofty sentiments, but a distinct feeling of celebration was in the air—and for good reason. To this day, this effort remains one of humankind's greatest feats of scientific exploration. Hundreds of researchers in numerous countries had collaborated with a single purpose. They had finally succeeded in identifying and precisely mapping the "language of life"—that is, the entirety of the genetic material of the human organism, or the actual chemical sequence of the more than 20,000 genes that make up the famed "double helix," the two twisted strands that make up each DNA molecule. Most of us believed at that time that acquiring this knowledge would be a boon to medicine—and ultimately to every person's prospect for health and longevity—because our genes contain the biochemical instructions that direct every aspect of our biological function at each moment of our lives.

As a result, we were told, a great revolution had arrived that would change the way my colleagues and I would practice. An "era of personalized medicine" would unfold before our eyes. Soon, the detailed knowledge of each person's genome would become an intricate part of our treatment plans. This virtual instruction manual would tell us doctors just what medications our patients should take, which diseases they were likely to get, and even how long they might live.

Along with this promise of the detailed understanding of the genetic origin of diseases came the new hope that, over time, our leading medical researchers would find ways to remedy the negative genetic inheritance or even harmful random mutations that might be found in a given individual's genome. In the first few years that desirable prospect seemed like a distant vision. But new technologies for the massive, parallel-processing of genomes had become a reality by 2005. It suddenly seemed like the future was already here.

Yet, more than a decade has passed since those heady days, and we have not seen the widespread application of advanced genetic technology in the ways that were promised. As new treatments trickled out, we slowly realized that only a tiny percentage of diseases could be directly treated with genetic interventions. Had something gone wrong with the glowing prognosis we were given for the future of medicine? Was our celebratory mood in 2001 premature or perhaps in vain?

In actual fact, the science of genetics was now poised to take a surprising and even more profound new turn. Gradually, an exhilarating new reality began to reveal itself, one that is now culminating in a new paradigm for human genetics and for the genetics of all living things. The successful mapping of the human genome was only a first big step, it turns out, one that would become the foundation for yet another quantum leap in biology. More advanced research, especially in the last decade, points to the advent of a new field called *epigenetics*, which studies the *human epigenome.*

One reigning assumption of genomic researchers had long been that our genes are deterministic in their expression. But the new epigenetic research demonstrated that this premise is largely false. Instead, we have come to see that our genes respond, or more specifically our epigenome responds, to *how we interact with our world.*

In other words, our lifestyle choices and our life conditions play a large part in how our epigenome functions, which in turn determines the function of the underlying genome. What we eat, what we drink and breathe, our stress levels, our use of pharmaceuticals, our interaction with the immediate physical and social environment—these are the essential factors in genetic expression. Most notably, studies in identical twins have conclusively shown that while each twin's genome remains identical throughout their lives, their individual epigenomes can vary significantly over time. And this variability can only be explained by the differing ways each twin lives out his or her life on a daily basis, as well as their environmental exposures.

Today we know that surrounding every gene is a complex set of switches that determine what property of that gene will or will not be expressed. These switches interact with—and are directly affected by—all of our lifestyle choices. As I will explain in this book, this new concept represents a major breakthrough in understanding the direct influence each of us can have on our inherited genes. Whereas we once thought our genes rigidly determine our biological reality, we now know that it is largely the other way around!

What can all these dramatic developments in the world of science mean for your life and your health? Actually, the answer to this question is almost stark. The upshot is that that our diet, exercise, stress management, and other lifestyle and environmental choices matter even more than before. In other worlds, if we change our lives, we *can* change our genetic expression.

And, you may wonder, are there real-world applications of this knowledge that you and your doctor can use today? Indeed there are. I will introduce to you specific medical protocols for applying the epigenetic approach to our biology right now, today, to improve your life and your health. They largely involve "reading" the results of state-of-the-art biological assays that involve at least three key components:

1. A genetic profile of the key genes that govern all of the major chronic diseases.

2. A comprehensive blood draw that will depict the hundreds of biomarkers in your blood that predict your state of health or illness with precision.

3. A "biomic" or intestinal tract assessment that tells you how well your body is actually using the nutrients you are ingesting as well as the impact of stress, exercise, and pharmaceuticals.

Together, these three sets of data—along with other indicators—provide an unprecedented, accurate, and even self-administered means of tracking the subtle impact of lifestyle changes that can move you toward optimal health and longevity. Because they set the stage for these advances, we should be grateful to the pioneers in genetics whose heroic work presaged this revolutionary new approach to health and medicine.

This book is laid out in six chapters that bring these and many other important lessons home.

In Chapter 1, "New Reasons to Hope," I survey the new era of medicine that is arising because of what we've learned about the human epigenome. It is true that certain rare and unpreventable diseases are caused by a single defective gene; but aside from these unusual cases, we now understand that scientists can only predict genetic *probabilities*, not certainties, and that our lifestyle choices are the crucial factor in shaping these probabilities into outcomes. Generally, we can identify inherited genetic proclivities for specific conditions like heart disease or type 2 diabetes, but the success of such predictions remains limited and contingent. Just as important, we have discovered that *genes don't work alone.* We sometimes have effective treatments for conditions governed by a single gene, but the more common disorders are the result of complex interactions between many genes and numerous lifestyle and environmental factors, as these are mediated by the individual's ever-changing epigenome. The complexities involved are catalyzing a new world of "big data" approaches to medicine that are sometimes called *bioinformatics* or *computational biology.*

These things are all important, but my most essential message is our ability to create our own biological reality and to determine by our actions our own heath and longevity. Ultimately, the new knowledge that we are active participants in determining our own prognosis for health is a profoundly spiritual message of hope.

In Chapter 2, "Keys to Wellness," I introduce the *biomarkers* that provide clues to our health prospects. We are beginning to list the markers that are the most essential indicators of health and disease. As noted, we will soon have accurate and inexpensive tests for these biomarkers that can help us determine how particular health practices and medical interventions are affecting our genetic expression. The good news is that an emerging panel of essential biomarkers, along with other criteria such as a "biomic" (intestinal tract) assay and a complete blood test, is becoming available now at a reasonable price. What's more, we may soon get the key list of markers down to only a few dozen in number. In other words, doctors of the future will not need to analyze your entire genome; they will only need to examine a handful of genetic and biological markers with the most influence.

With your profile in hand, doctors will soon be able to design a highly personalized health plan that covers all the bases: diet, drugs, exercise, meditation, stress management, psychosocial and environmental influences, as well as other lifestyle recommendations. Then, at the end of a ten-to-twelve-week period of carrying out these recommendations, we can expect that the genetic markers we are measuring should improve. We simply need to retest them again, and continue to make adjustments at regular intervals in light of the findings of the most recent test. With the advent of this *personalized medicine* approach that is based on epigenetics, we can focus on very specific, individualized behaviors, practices, products, and services that optimize your health.

In Chapter 3, "Epigenesis," I explicate *seven crucial biological pathways* in the body that my colleagues and I have identified, and then I explain in brief how to facilitate optimal wellness along each pathway. Inflammation is one of these key pathways; and we now know, for example, that extreme psychological states—such as trauma, depression, and even job stress—can switch on an inflammatory response in certain genes. Plus, certain high-fat diets can switch on the expression of these same genes, which in turn increases inflammation in the arteries, a major risk factor for heart disease. In Chapter 3 and the chapters that follow, I detail specific changes in lifestyle that correspond with each of the pathways. In brief, these seven gateways to health include:

1. *Oxidative Stress*, which occurs when the body is unable to eliminate the damaging by-products of excessive oxygenation.

2. *Inflammation*, which is a major risk factor for virtually every chronic disease, but fortunately is easily influenced through nutrition and stress management.

3. *Immunity*, a crucial pathway that protects us from infection caused by external agents.

4. *Detoxification*, which is critical in every cell throughout our bodies.

5. *Lipid Metabolism*, a form of fat metabolism, which is a pathway that responds well to proper dietary changes.

6. *Mineral Metabolism,* which concerns biomarkers that help us identify how well our bodies are metabolizing minerals.

7. *Methylation,* the best known of the epigenetic switching processes that control DNA expression.

With the essential knowledge of the seven biochemical pathways in hand, we can identify patterns that point to more specific steps you can take to influence these passageways. I present what we know about these patterns in the final three chapters.

Chapter 4, "Nutrigenomics," homes in on the growing number of studies that have shown that there are vital nutrients or plant compounds that can "talk" to our genes, turning genetic messages on or off. In this chapter, you will especially learn how to maximize your health by eliminating metabolic syndrome—that cluster of conditions that increases the risk of heart disease, stroke, and diabetes.

We all wonder which diet optimizes our health in the face of a confusing flurry of often contradictory options. Fortunately, more advanced studies, often based on genetic analysis, will remove much of this speculation. It will provide a more objective, scientific basis for the optimal diet for each person. This chapter will make specific dietary recommendations, describing their genetic impact on the key pathways and biomarkers. It also covers today's increasingly important research into the *gut microbiome*, which is looming large as a major factor in health and disease—and even in epigenetic regulation.

Chapter 5, "Mind Matters," explains how to turn off our genetic vulnerabilities both by reducing stress and by developing emotional well-being. My first book back in 1977 provided scientific proof of the link between stress and the major types of disease, and offered compelling evidence that the mind holds a profound influence over the body. That insight has remained a theme throughout my research, clinical practice, and writing ever since.

Conclusive evidence makes it clear that our beliefs, attitudes, and emotions—be they positive or negative—have a direct, causal, and enduring impact on the DNA core of every cell in our bodies. For example, we have compelling evidence that the experience of childhood trauma negatively influences a young person's capacity to respond to ordinary stress as he or she reaches adulthood. Epigenetic influences literally burn trauma into the brain cells of a child. However, just as difficult events can affect our genes, so is it possible to have a *positive* impact on the expression of our genetic code in childhood as well as adulthood.

The concluding section, Chapter 6, "The Era of Personalized Medicine," provides a broad look into what the future holds for self-care, medical care, and national healthcare. Our goal should be to embrace the new world described in this book and dramatically accelerate biomedical progress in the light of the epigenetics revolution and the allied advances in computational biology.

Even today we have the advanced technology and know-how that is required to get started in earnest with truly personalized medicine. It is now only a matter of making the techniques discussed in this book more readily available in laboratories across the country, then educating both patients and doctors to apply it to their everyday lives. We are at a point when literally anyone can self-administer a simple, inexpensive set of assays to determine exactly what is needed for optimal heath and longevity. Once this data is analyzed, sometimes with the use of artificial intelligence software, there will be much less guessing about the "right" diet, or which supplements or herbs to take or not, or whether a particular exercise is beneficial, or how effective your stress management practice or your medication is—or any other health question or issue. Readouts, some of them in real time, will tell you almost immediately what the effect a given health practice or protocol will have on your body-mind system.

The futuristic methodologies I present in the closing chapter are already proving to be more effective than the current reactive, episodic, fragmented, and impersonal treatment of disease episodes. That's why I argue in this book that it is time for a seismic shift in the way we deliver medicine.

Ultimately, the same tools that are introducing us to personalized medicine will lead to new approaches in prevention, so we can eliminate more conditions that lead to disease before they start. But when early preventive measures or when later lifestyle changes are not adequate, we can justifiably turn to our physicians to come up with the right drug treatment at the right dose at the right time, with minimum side effects, and maximum effectiveness—thanks to the epigenetic tools they will have in hand.

At the center of this brave new world of health care will stand *you*—your commitment to be educated and engaged in your own health in a way that is both cost-effective and a sustainable boost to our national health and economy. Most importantly, the advent of personalized medicine based on the new findings of epigenetics heralds an age of optimal health that is unparalleled in human history. Because of these advances in science and technology, a hopeful future is just up ahead—as you will see in the coming chapters. This truly *is* cause for celebration.

1

NEW REASONS TO HOPE

What We Have Learned about Our Genes

In May 2013, celebrity actress Angelina Jolie suddenly made headlines—not because she was starring in a new movie but because she had made a drastic medical decision. She announced to the world in a *New York Times* op-ed that she had undergone a preventive double mastectomy.[1] Her decision to submit to such invasive surgery was, she wrote, the result of genetic tests that indicated she had an 87 percent likelihood of developing breast cancer. Jolie also shared that she had lost both her mother and grandmother to ovarian cancer, which is closely tied to breast cancer. As the mother of six young children, Jolie decided on their

1 Angelina Jolie, "My Medical Choice," https://www.nytimes.com/2013/05/14/opinion/my-medical-choice.html (May 14, 2013).

behalf "to be proactive and to minimize the risk as much as I could."[2]

Jolie's dramatic story had worldwide impact. The *New York Times* later reported on how awareness of the breast cancer issue had exploded in Israel as a result of Jolie's announcement. Jolie is of Ashkenazi Jewish descent, and it is known that about half the Ashkenazi women in Israel and the majority of them in the United States are likely to have the same mutation as Jolie. As these susceptible women learned of their increased genetic risk, the *Times* noted that they face the same crisis that Jolie did before them.[3]

Was the famed actress misguided in making such a drastic choice? Or is preventive mastectomy the only ethical and reasonable medical choice for women with Jolie's mutation? If the answer to the latter question is yes, should insurers cover this procedure for all women with this mutation?

Giving informed answers to questions like these requires a short course in the revolutionary new understanding of human biology that we examine in this book. The discovery early in this century of the significance of *epigenetics*—and more recently of the genetic role of our body's symbiotic cousin, the *gut microbiome*—was the tip-off to researchers that something far more complex was going on than had ever been imagined

---

2 Dr. Andrew Weil, "Did Angelina Jolie Do the Right Thing?," https://www.drweil.com/health-wellness/body-mind-spirit/cancer/did-angelina-jolie-do-the-right-thing/ (May 7, 2013).

3 Roni Caryn Rabin, "In Israel, a Push to Screen for Cancer Gene Leaves Many Conflict," http://www.nytimes.com/2013/11/27/health/in-israel-a-push-to-screen-for-cancer-gene-leaves-many conflicted.html?_r=0 (Nov. 26, 2013).

in twentieth-century medicine. Because of these developments, we can now safely say that Angelia Jolie had unwittingly based her decision on a model of genetic determinism that is on the brink of extinction—as you will soon learn.

The End Game for Genetic Determinism

Advances in genetics of the last few decades have been nothing short of astonishing. Scientists can now locate and map (or "sequence") every single gene in the human genome at an increasingly lower cost. The relative ease and precision of today's *gene-sequencing technology* now makes it commercially feasible to identify the biochemical makeup of every one of the over 20,000 genes in each person's DNA, as well as every other molecular feature of the DNA strand on their inherited genome.

The first effort at comprehensive gene sequencing in 2001, the Human Genome Project, cost about $100 million. Fifteen years later, the cost had dropped to about $500, and continues to fall from there. Impressively, these techniques allow us to zero in on the tiniest molecular units of our DNA, known as *base pairs.* Geneticists designate these biochemical units by a simple series of four letters. Sequencing techniques allow them to locate those genes that carry potentially harmful *mutations*—rogue base pairs whose letters are out of proper order. These unique variations in a standard base pair are also known as *gene variants.*

Is there a practical import for your health? Yes, indeed there is, because gene variants can often be correlated with some degree of risk for a specific disease. Such a statistical association of a variant with a particular disease makes you vulnerable to it; only rarely is it a certainty. The new science of epigenetics shows that many other crucial influences are at work in creating disease conditions in the body that may or may not activate this genetic predisposition. Plus, some variants can actually be beneficial adaptations—though again, rarely.

Because advances in gene-sequencing now make it easy to locate variants, a giant new industry for gene mapping is beginning to have wide influence in clinical medicine. And while it is a wonderful gift to become aware of our inherited genetic traits and possible disease vulnerabilities, putting such information in the wrong hands can also be dangerous and misleading.

According to the National Institutes of Health, the genetic testing industry will grow to about $20 billion by 2020. Most of these billions will be spent on promises to predict your risk of major diseases. But what the public doesn't know is that such genetic tests can predict with certainty only a *few percent* of all known diseases. All other cases of disease occurrence depend at least in part on factors outside your inherited genome, most notably your lifestyle and your particular life conditions.

Nevertheless, because of the persistence of the mechanistic paradigm of genetics inherited from the last two centuries, massive resources are being spent on predicting genetic diseases and matching drugs to such conditions. This can be a blessing to those millions of Americans who suffer from the inexorable expression of one of roughly 5,000 rare genetic diseases that afflict about ten percent of the population. But what about the vast majority of us who don't have such defective genes? An even greater blessing would be to refocus genetic research on optimizing the expression of our *good* genes! This book reveals the state-of-the-art of this more expansive and proactive approach to human biology.

The difficult case of Angelina Jolie illustrates the transitional state of the genetics industry, as we wait for clinical practice to catch up with the new biology. Jolie has a well-understood mutation in genes known as BRCA1 and BRCA2, which work as tumor suppressors. This relatively common mutation can make these genes incapable of performing their important function, giving women with these variants a high risk of both breast and ovarian cancer.[4]

Brace yourself, because what follows is a graphic description of the aggressive intervention necessitated by Jolie's decision. Once she knew her test results, Ms. Jolie opted for a complex form of preventive surgery that requires three consecutive operations over several months. First, she underwent a procedure designed to spare the nipple and surrounding areola.

---

4 This fact poses a difficult challenge for these susceptible women, but the presence of the BRCA mutation isn't their only consideration. A history of breast or ovarian cancer in the family is also a factor, with or without the BRCA mutation. Women are more likely to inherit this harmful mutation if close relatives have had these cancers, as is the case with Jolie's family.

Next, surgeons removed all the breast tissue while saving the skin that contains the breasts. In a third procedure, her breasts were reconstructed with implants. This procedure only targeted her breasts because, as she wrote, "my risk of breast cancer is higher than my risk of ovarian cancer." Jolie still has to face a decision about preventive surgery on her ovaries as well—yet another drastic and expensive intervention.

The Better Choice: *Change Your Gene Expression*

Without a doubt, Jolie faced a big risk for breast cancer and made a tough decision. But we can only wonder to what extent she was made aware of the valid alternative approaches for women who face this dilemma, even within the old paradigm of medicine. For example, a far less invasive approach would have been to take *tamoxifen*, an estrogen-blocking drug. Or, she could have opted for preventive medical monitoring in an effort to catch breast cancer early. The best approach of all would have been to combine these two treatments with what we in California call a "reframe"—that is, to simply leave behind the outmoded concept of genetic determinism and embrace the science of epigenetic modification. This would entail that she consciously change her patterns of gene expression in ways that *compensate* for her inherited BRCA1 and BRCA2 mutation. We now know that the most powerful option for Jolie—or anyone with a known genetic predisposition to any disease—is to change their environment and their lifestyle in the ways we will discuss in future chapters. Such an approach can *reduce or may even eliminate* every type of inherited genetic risk, with the exception of those rare genetic diseases that are virtually irreversible.

And you and I can go even further with this new understanding: We can engage in practices that *optimize gene expression for a lifetime of sustained wellness.*

We sometimes designate this new approach to human biology by the term *epigenesis*, which conveys the sense that our genes and DNA are dynamic and fluid in their expression. Evidence for epigenesis is now very well-established in the leading medical journals. Hundreds of studies show that our genes are responsive to the biochemical and energetic environment we create in and around our cells through our daily choices. As a result, a thrilling new picture is emerging: the discovery that our biology is consciously modifiable. Scientists are discovering that our bodies—and our gene expressions—quickly adapt to new conditions, and they are also learning that these adaptations can be traced to specific *biomarkers* (covered in the next chapter) that you and I can target to boost our health prospects. The general academic discipline concerned with this approach is *systems biology*, and its clinical application has been called *functional medicine.*

Yet, the mind-sets of many geneticists and doctors are out of step with these newly discovered realities, too often because of biases in favor of the old paradigm based on previous training—or sadly, because they are driven by the momentum of commercial considerations. As a result, today's biomedical science is riddled by at least two very divergent approaches. Genetic researchers and medical clinicians are diverging into two general types:

- **The mechanistic approach:** Those who focus on using whole-genome mapping to identify mutations that have a probability of resulting in disease, with the aim of developing drugs or surgical procedures that either treat these genetic predispositions preventively or after the associated genetic disease appears.
- **The systems approach:** Those who search for *modifiable biomarkers*—including gene variants, epigenetic modifiers, and biomic markers in the gut—and who use this information to design a set of lifestyle and environmental changes that create measurable health improvements in the targeted biomarkers.

Clearly, the first approach above is of course the one chosen by Ms. Jolie. This reductionistic understanding of the genome has given rise to a host of companies that exploit the fear that our genes dictate our destiny. Again, it is true that inheriting certain gene variants guarantees you will get a rare disease; but the overemphasis on that isolated fact has gone so far that, as we'll see, government regulators have had to step in powerfully to protect the public.

This lesson especially applies to all of us concerned with optimal health, especially if we work in the healthcare industry. All of us will need to modify our business practices *and* our health practices in light of the research that conclusively proves that our genes respond—or more specifically our epigenome and our gut biome respond—to how we interact with ourselves, with each other, and with our world.

Introducing the New Science of Epigenetics

We've noted that when the map of the human genome was first revealed, it was believed that geneticists would soon be able to make solid predictions about which diseases each of us would get as we age. But the immediate aftermath of the mapping of the human genome has led in a radically different direction as we begin to understand the human epigenome. We now know that environmental and psychosocial factors as well as lifestyle choices play the largest part in how our epigenome functions, which in turn determines the expression of the genes that govern our health and longevity.

Epigenomics, the new scientific discipline of research into the epigenome, is the study of the *chemical tags* that park themselves on the genome that literally control the activities of our genes. In a sense, these markers appear "above" the genes—and is thus signified by the Greek prefix "epi," which means "above" or "upon." It is almost as if there are two languages being "spoken" by our DNA: the original "script" of our genome, and a secondary and more powerful linguistic control system that sits on top of each gene. This system determines, more than 95 percent of the time, whether, when, and how much a given gene (or some other portion of the DNA strand) is permitted to express itself as it does its routine work of "coding" for a myriad of biochemical activities in the cell.

The analogy of a theatrical script helps illustrate how epigenetic regulation works. Perhaps you have seen different movie versions of William Shakespeare's *Hamlet*—for example, those featuring Richard Burton (1964), Kenneth Branagh (1996), or more recently Benedict Cumberbatch (2015). These films differ greatly from one another, but of course the words on the page in Shakespeare's underlying script never change. Shakespeare's original theatrical script can be compared to our genetic code, and the differing performances of *Hamlet* are analogous to the function of epigenetic regulation. Once we inherit our unique genome, how it appears "on the printed page" remains stable throughout our lives; what is critical is our *expression* of these "lines" of the DNA script in and by the way we live our lives. In that sense, we are like actors who are "directing" the "performance" of our genetic script in ways that are unique to us, but in a manner that is also conditioned by the "theatrical set"—our immediate environment.

Allow me to extend the metaphor a bit more. If we could read our epigenome, it would look like a director's "notes" that we have written above the lines or in the margins of our genetic "screenplay"; this "shooting script" provides the epigenetic modifications that are made possible by our directorial choices. In essence, we direct our biological lives in the same sense that a movie director determines the expression of the underlying script by the actors (not to mention set designers, lighting and camera crew, etc.).

Bear in mind, however, that the variations in gene expression that make up our epigenome do not govern *every* biological trait or function. Certain physical characteristics, such as eye color or height, are one hundred percent predetermined by the inherited genes. This characteristic of gene expression is known as *gene penetrance.* For example, in identical twins, the penetrance of their genes for physical appearance is one hundred percent guaranteed. Plus, as noted, gene mutations that program for certain rare diseases also have this level of predictability—they can't be modified epigenetically. But again, complete penetrance is the exception, and not the rule—as we will soon observe in living color when we examine the genetic studies of identical twins.

Penetrant genes make up only about five percent of the human genome. In all other cases, genes for thousands of functions must be either activated, suppressed, or modified by epigenetic mechanisms. That's the bottom line for this discussion, but there's one more feature of the epigenome to ponder later on. The epigenetic alterations that you may acquire don't just change your biology during your lifetime; some of these modifications can be passed on to future generations that follow you. This surprising phenomenon is known as *transgenerational epigenetic inheritance.*

If I may use yet another analogy, we can compare epigenetic regulation to switching on or off a light in your bedroom.

Much like our genetic code, the light switch on your wall and the light bulbs and their fixtures that are connected by wires to this switch are a stable presence—they are the infrastructure that always remains in place. But *you* are the determining factor in this simple equation. You must decide whether or not you want the light to shine, or whether this light should be turned up or down in intensity.

Our growth from the time of our conception and our daily health practices and habits—along with the routine moment-to-moment shifts in the functions of the tissues and organs of our bodies through a variety of biochemical pathways—all these factors constitute a galaxy of biological changes that determine our well-being in this life. All of these elements are orchestrated by the biochemical switches that constitute the epigenomic superstructure that sits on top of our inherited genetic infrastructure. Specific chemical reactions are able to *switch* the relevant parts of our genome on and off or up and down at strategic times and locations on a given gene or on the vast regions of the DNA strand that were once known as "junk DNA." Epigenomics is the study of the biochemistry that regulates those switches—for humans as well as for all organisms.

By way of illustrating this vital principle of gene switching, let's briefly consider one of many determining factors that have been closely studied: the impact of nutrition on the epigenetic modification. What you eat, whether that consists of healthy nutrients or processed chemicals, provides the chemical environment in which your cells are sustained and in which they bathe. We now know that eating a nutritious diet high in fresh vegetables can add a layer of protection to your epigenome; that's because certain compounds the body uses to switch off harmful genes are uniquely found in your veggies. In this case, it's a matter of providing your body with the basic chemical building blocks of the epigenetic switches themselves. If these chemicals are present, they can ensure that all the switches are turned off that actually need to be off in order to ensure healthy functioning. For example, if your diet is low in *folate* (a component of the B-vitamin complex), some harmful genes may be left on because there wasn't enough "off factor" to go around.

In other words, your genetic expression is determined, in part, from what *you* decide to eat—which in turn programs the function of biochemical regulators in your epigenome. As yet, not many hard-and-fast studies detail exactly what foods or supplements, or the lack thereof, will lead in a straight line to a specific disease or health improvement. But there is a large amount of evidence that what you are eating does indeed condition your genetic expression in innumerable ways. Details of the known dietary influences will be spelled out in Chapter 4.

But again, always bear in mind that this example of our diet's impact on our genes also holds true for exercise, environmental toxins, stress, emotional states, smoking, radiation exposure, pharmaceuticals, and many other influences that we are now beginning to map and understand. And if this sounds like we have come a long way from the deterministic paradigm that led to Angelina Jolie's decision, you would be correct.

One of the many discoveries that led the way to the epigenetic revolution were studies in identical twins, the results of which have conclusively shown that while each twin's underlying genome remains identical throughout their lives, their individual epigenomes can vary significantly over time. And this variability can only be explained by the *differing ways each twin lives out his or her life on a daily basis.* Perhaps the most important study of identical twins was carried out at Johns Hopkins University, which we now examine.

Game Changer: Studies of Identical Twins

Identical twins have identical genomes—a marvelous fact of nature. But we now know that twins vary greatly in their incidence of disease and other life changes. To get to this weighty finding, genetic researchers have asked one key question: If one twin gets a disease, what are the chances that the other twin contracts this same disease, say, in a fifteen-year period? The results are eye-popping. Studies have shown that, for example, their chance of getting Parkinson's disease is *only five percent*; this means that if one twin comes down with it, the other twin's lifestyle and environment plays a 95 percent role in determining whether they too will contract Parkinson's. With respect to coronary heart disease, the likelihood is 50 percent—no more than random chance. For most cancers, the chance the other twin will suffer from this disease is less than 50 percent. You can see that a remarkable statistical trend shows up in these studies.

In 2012, Dr. Bert Vogelstein of Johns Hopkins University Medical School announced the results of perhaps the most comprehensive genetic study of twins ever done.[5] He and his team compared the genomes of thousands of identical twins and confirmed—this time by referencing a very large data set—that disease cannot be predicted by genes alone other than in exceptional cases. They also concluded that the medical use of whole-genome sequencing to determine our genetic susceptibility to specific diseases can not only lack significance but can even be misleading.

In this breakthrough project, the Johns Hopkins team focused on 24 major diseases. Their complex analysis showed that genetic tests could alert most individuals to an increased risk of only *one* disease on average, which means that such tests are a rather poor predictor. In other words, a profile of a given twin's whole genome in comparison to that of their twin would give a negative test outcome for the vast majority of diseases known to humankind. But such a finding does not mean anything like "free pass," points out Vogelstein and his team. This individual will still have the same risk for all other diseases as the general population.

---

5 Nicholas J. Roberts, Joshua T. Vogelstein et al., "The Predictive Capacity of Personal Genome Sequencing," *Science Translational Medicine*, https://www.ncbi.nlm.nih.gov/pmc/articles/PMC3741669/ (May 9, 2012). See also "Whole Genome Sequencing Not Informative for All," Johns Hopkins Medical School News, https://www.hopkinsmedicine.org/news/media/releases/whole_genome_sequencing_not_informative_for_all (Apr. 2, 2012).

Indeed, each member of a set of identical twins will get all sorts of diseases according to the general statistical risk—just not the *same* diseases as the other twin!

The genetic study of identical twins proves once again that, with some exceptions, our genes do not inexorably determine our fate, but instead adapt to how we direct our daily lives. And when you consider that evolution favors adaptation, this finding makes perfect sense. From the very beginning, our survival as a species has depended on our ability to adjust biologically to relentless change. Studies of twins make it clear that directing our lives in such a way that we adapt to changing life conditions is built right into our genetic makeup, showing up there as the director's ever-changing "shooting script," the epigenome.

More Basics of Today's Genetics

Genetic sequencing has come a very long way, but the basic technology that researchers currently use to map the genome has been in place for over thirty years. The method for sequencing was invented in the mid-1970s by Dr. Frederick Sanger of the Medical Research Council in Cambridge, England. His first step was to isolate the long strings of double-stranded DNA from cells. Sanger and his successors soon discovered that attached to each string are millions and millions of tiny chemicals that we noted before now are called *bases*, of which there are four types. These four fundamental chemical units go by the names of *adenine*, *thymine*, *guanine*, and *cytosine*, and researchers refer to these bases by their initials, A, T, G, and C. Different permutations of these letters form the millions of rungs on the ladder of the DNA helix, and in each rung we find that an A is always linked with a T, and a G is always paired with a C—which is why they are known as base pairs. This little sketch of our DNA's structure may sound simple, but what complicates the work of sequencing the entire human genome is that each gene is made up of more than 3 billion pairs of these bases from each of your parents. When such base pairs are linked to either side of that living ladder—the double helix that makes up a DNA strand--that molecular unit is known as a *nucleotide*; in other words, each nucleotide is joined "at the hip" to each side of the helix. Large sets of these DNA strands are, in turn, lumped together into the 46 *chromosomes* found in each cell that you learned about in high school biology.

We should note here that a very tiny percentage of our inherited 3 billion base pairs, roughly 3 million of them, are the pesky gene variants I introduced earlier. These are *mutant* nucleotides. They are the culprits that can sometimes guarantee a certain disease. Far more often they simply make you susceptible to the expression of a certain disease or may predispose you to a certain behavior or personality trait.

Each gene is a basic functional unit of the language of the genome. Genes spell out a code—much like letters in the alphabet spell out words—and the basic unit of this genetic alphabet is the four-letter base. But how do mutations actually occur in this alphabet? Due to a wide variety of biochemical factors, single letters can get deleted, added, or rearranged, resulting in a mutation or variation. We noted that you may inherit your parent's gene mispellings as part of your genome, but you also generate unique mutations in the course of the biological spelling bee of your own life. However it may occur, these altered genes may continue functioning normally if the mutation is minor, because the epigenome has built-in safeguards that can repair mutated genes.[6] But sometimes even this "back up" system breaks down. In theory at least, a disease like cancer can be the result of such a double failure.

Gene expressions are almost always mediated by our epigenome. You may inherit a strange gene or generate a mutant nucleotide that codes for a specific disease, but again, if this unit of your DNA stays in the off position, the change it codes for will never express itself; on the other hand, environmental or lifestyle factors may turn this gene on. In practice, however, bear in mind that gene regulation is actually far more complex than a simple binary difference. Many genes work on what might be called a sliding scale, more like the rheostat switch you use to turn your dining room light up or down in intensity. In addition, genes communicate fluidly with one another. Some genetically determined characteristics require

---

6 For example, using our alphabet comparison, a mutation may cause "dog" to become "ddog," but this sort of genetic typo is fairly harmless because it is close enough to the original "meaning." Or, let's say it became "hog." This is still okay, because a hog is not a dog but at least it's a four-legged animal! But if the spelling becomes "ddgg," serious difficulties may result. That gene may begin coding for something else entirely new that cannot be predicted. Or, an important gene may be simply rendered nonfunctional.

only a single gene (and thus are penetrant), but the vast number of genetic influences are created by a dance or symphony of genes delivering dynamic input to the whole system, along with the many epigenetic influences we have identified.

Ultimately, the lessons of epigenetics mean that control is being given back to you and me. We are no longer puppets of our DNA. Our human genome is like the theatrical stage for the future steps in evolution that we ourselves direct, making the power of choice an integral part of genetics. Unless our personal decisions about lifestyle and environment are all taken into consideration, a full picture of the mysteries of our DNA can never be attained.

Genes Don't Work Alone

Because of today's revolution in our understanding of DNA, complexity is now the rule in biology and medicine. Geneticists have learned that even specific physical traits are not determined by single genes. For example, no one gene exists for height; more than 20 genes have been identified so far, and we know little about how their interactions might contribute toward determining how tall we will grow up to be. And while single gene mutations can explain some rare diseases, some of the more complex diseases such as breast cancer, heart disease, and even the majority of Parkinson's cases can be traced to unfortunate combinations of a set of *normal* genes.

We understand very well how a single gene codes the instructions (through its associated RNA, to be discussed later) to create a particular protein—which then turns on a certain trait or bodily process. But we know far less about the operation of *systems of genes*, including how the genome, epigenome, and gut biome ultimately work together as a whole to regulate the human organism, thereby enabling clinicians like myself to guide patients in the right direction. Some call this great systemic collaboration the *supergenome*.

Because of their training under the old paradigm, most doctors prefer to think in terms of simple cause and effect: symptom "X" can be traced to cause "Y." But the emerging story of the complexity of our biology points to a very different picture. We've noted that the cutting edge of academic biological research is now known as *systems biology*, which has led geneticists to think in terms of a "cloud" of influences leading to outcomes—good, bad, or ugly. Fortunately, most of these factors are potentially under our control, provided that we embrace the lessons of the epigenetic revolution and continue along the lines of research recommended in this book.

Like a real cloud, the "biological cloud" changes almost moment to moment. That's because our bodies are designed to flexibly adapt to inputs of every kind, including the biological impact of our thoughts and feelings. Because of the more inclusive perspective of mind-body medicine that I have championed for my whole career, the mind itself—and even the impulses of the human soul and spirit—are influences that add complexity to the genomic equation.

One of the pioneers in the exploration of biological complexity is Dr. Eric Schadt, Chairman of Genetics and Genomics Science at the Icahn School of Medicine at Mount Sinai. Dr. Schadt and his colleagues have concluded—like all researchers on the cutting edge of this field—that most diseases cannot be explained by the conventional mechanistic view of genetics. Instead, he believes, illnesses (and good health) are caused by a vast network of biological influences that can be understood only with advanced computer modeling.

Because he is an articulate exponent of the medical use of *big data* or *bioinformatics*, Schadt has even been featured in popular magazines such as *Wired* and *Esquire*. As the story goes, at one point in his career he headed a research lab at Merck (a large American pharmaceutical company) that was tasked with developing new drugs. Schadt's work was considered a great success by the company, but he soon became disillusioned. His team's research taught him that the underlying premise of conventional biology was false—that is, the idea that we could understand disease one gene at a time and target drugs at defective genes. Schadt eventually told Merck that "this was a strategy doomed to fail, because disease arose not from single genes or pathways but rather out of vast networks of genes and pathways whose interactions could be understood only by supercomputers guided by abstruse algorithms."[7]

As researchers have faced up to this complexity, they have turned to what is known as "genome-wide association studies"

7 Mark Warren and Tom Junod, "Patient Zero," *Esquire*, https://www.esquire.com/lifestyle/news/a23509/patient-zero-1213/ (Nov. 19, 2013).

(GWAS). This is one form of Schadt's big data approach. In such large-scale studies, the genetic sequences of many thousands of individuals with a specific disease are mapped and studied with algorithms in order to determine if there is a statistical correlation between a particular gene or set of genes and a specific condition—across this huge sample of subjects.

As this more advanced and data-intensive research progresses, it is still essential to acknowledge the progress we have made with single-gene or *monogenic* maladies that are 100 percent penetrant. One instance where genetic testing is accurate and predictive is for Huntington's disease, a very unfortunate inherited condition in which nerve cells in the brain break down over time. A gene for this disease was first identified in 1993, in the world's earliest attempt to locate a single disease-causing gene by tracking variants in the human genome.[8] A genetic test for Huntington's is now widely available, but for those individuals at risk for inheriting it, the decision as to whether to get tested or not can be excruciatingly painful, since there is no known cure for this disease.

Exceptions to the rule, such as the case of Huntington's, actually prove our rule. They serve to call our attention to the role of the epigenome and ultimately to the role of human consciousness in guiding our health decisions.

---

8 The co-discoverer of this gene is Rudolph E. Tanzi, PhD, who is the coauthor with Deepak Chopra of an insightful book on epigenetics called *Super Genes: Unlock the Astonishing Power of Your DNA for Optimum Health and Well-Being* (Harmony Books, 2015). Tanzi and Chopra also coined the term *supergenome*, used earlier.

Genes Are Also Governed by Our Beliefs and Choices

You may remember this oft-quoted line from Shakespeare's Hamlet, "The fault lies not in the stars but in ourselves." Allow me to paraphrase that quotation for our purposes: *The fault lies not in our genes but in ourselves.* The new era of epigenesis points us right back to our own habits and decisions. Our inherent power to choose our destiny is a spiritual reality that, in turn, governs our biological reality. We are not victims of genetic programming; instead, the truth is just the opposite.

If we and our doctors—and the scientists we fund with our tax dollars—believe that we are helpless pawns fated to live out our genetic code, then we will pour all of our research funds into discovering what is encrypted in our cells. However, the new knowledge that we are active participants in determining our own life directions is a profoundly spiritual message of hope and deserves at least as much scientific and medical attention.

There is a wonderful true story from John Lennon, who was asked about what he thought the purpose of his life was to be. He quipped that, "When I was five years old, my mother always told me that happiness was the key to life. When I went to school, they asked me what I wanted to be when I grew up. I wrote down 'happy.' They told me I didn't understand the assignment, and I told them they didn't understand life." For individuals seeking optimal health and longevity, John Lennon's response makes a great deal of sense. Choosing to be happy may in fact have a profound impact on how our genes are expressed in our lives.

Without question, our minds and emotions directly affect gene activity. And since the mind is the source of our lifestyle and behavior, it ultimately directs our biological destiny. Self-awareness holds the key to this process of transformation. Consciousness invisibly reaches into the biochemistry of every moment of life. In your body, as in every cell, the phenomenon of epigenetic regulation is integrative, self-generating, self-organizing, and self-directing in concert with the status of your beliefs and your commitment to conscious living. We'll turn in detail to this rich subject in Chapter 5.

Technology will also act as an ally to consciousness. Doctors are just now able to demonstrate the epigenetic effects of our conscious or unconscious daily choices with an advanced blood test, plus other assays of biomarkers. Many of these tests are already available but underutilized. Adding to this evolution is the development of wireless devices that can sense—in real time—many of our most basic biological functions, ranging from pedometers measuring our footsteps to heart monitors that relay changes in our heart rate and regularity. We can even expect "ingestible nanotechnology," a tiny microcomputer smaller than the head of a pin inside of a sugar pill. This nanocomputer will transmit signals on sleep patterns, stress hormones, blood sugar levels, and an array of basic biological functions. This data can be displayed to each individual, and it can be sent to heath care providers.

In the coming chapters you will learn that these technologies exist now, but the medical infrastructure that ranges from the training of doctors to the certification of labs will need time to catch up. That day is envisioned by Dr. Francis Collins, director of the National Institutes of Health (NIH) and one of the original leaders of the Human Genome Project. He has observed that "medicine for most of human history has been one size fits all. But we're all different, and the diseases we have lumped together under one label we're finding out are actually at the molecular level quite distinct.... Precision medicine tries to understand what's underneath those disease layers and tries not to lump everyone together but instead think about individual differences."[9]

That is not an idle fantasy but an evolving reality of the healthcare system of the near future. Doctors will not need to analyze your entire genome, which after all is only one factor among many. Instead, they will penetrate the "biological cloud" to discover the biomarkers with the most influence. In fact, we already know many of the markers that matter most to your health.

One day soon you will be able to see exactly how your gene expressions uniquely respond to any given choice or behavior. For reasons we don't yet full understand, some people will benefit greatly from eating a vegan diet, for instance, while others can be more flexible; or, for example, low-fat diets will be shown to be the healthiest choice by far for some

9 Quoted in Kate Tracy, "Bringing in a New Era of 'Precision Medicine,'" https://www.1776.vc/insights/weekly-trend-bringing-in-a-new-era-of-precision-medicine/ (Jan. 26, 2015).

people, but not for others. The biological assays of the near future will settle such mysteries. A set of markers that are modifiable by specific lifestyle practices will become evident within a span of time as short as a few days to an outside maximum of 10 to 12 weeks. Determining, changing, and monitoring our epigenetic profile for optimal health and longevity will become a widespread practice.

We may not yet have definitive studies showing the impact of every element of our lifestyles and environment, but we do have an enormous amount of evidence. In the chapters that follow, you will learn specific dietary and stress management practices that can literally change the expression of your inherited genetic code, prevent disease, and optimize your health. You don't need to wait for an imagined destiny to befall you. Your beliefs and choices are determining your wellness and longevity *right now.*

2

KEYS TO WELLNESS

Biological Markers That Govern Your Health

Since the epic moment when the first sequencing of the human genome was announced, genetic research has become exponentially faster, cheaper, and more revealing. In fact, decoding complete genomes has now become routine for today's genetic researchers, laying bare an embarrassment of riches in genomic data. So, you may wonder, does it make sense to have your own personal genome mapped in an effort to find markers for inherited diseases? Or, could a complete genetic profile provide indicators that will help you optimize your health? We've made it to first base in answering such questions, so in this chapter we'll go deep to center field for more specifics.

So far we've learned that whole-genome mapping for identical twins is not very predictive for the health issues that will befall them later in life. Instead, factors such as diet, exercise, exposures to toxins, and the balance of bacteria in our gut biome have much more bearing on your health prospects. On the other hand, gene tests can tell us if we have inherited an irreversible rare disease, and they can alert us about genetic vulnerabilities that *can* be modified through correct health practices or proper medical interventions. But then again, some diseases don't even result from genes we can map; for example, cancer often arises from mutations that can form at any time. In other words, not discovering a cancer gene in a genome test doesn't mean that you won't develop this disease later on because of your choices.

Since these points have been made earlier, you may wonder why I seem to harp on this issue. I do so because of the all-pervading presence of the mechanistic, "disease-care" paradigm of medicine. In the face of the profound findings of the epigenetics revolution that I report on in this book, this older model of human biology still remains dominant in medical schools, drug companies, regulatory agencies, most other scientific establishments, and in the mass media. As a result, too much genetic research is focused on identifying gene variations that may be linked to particular diseases. This data can be useful, of course, but such knowledge cuts both ways. As we saw in the case of Angelina Jolie, gene-test results can be misleading if treated in isolation from the more essential factors. And among these factors is the discovery that we shape our *epigenome* each day by our decisions, and further, that this sensitive and malleable biological "script" governs the vast number of our health outcomes. In addition, our personal epigenesis leaves behind crucial biomarkers we can detect and modify—a key topic of this chapter.

Incidentally, I've just described the precision medicine of the future. For now, the medical establishment will continue to favor the simplicity of gene hunting, perhaps because of our cultural bias toward the technological quick-fix. But if we combine this lazy attitude with the ravages of today's biological reductionism, much can go wrong. An inevitable result will be the overdetection and overtreatment of "genetic" diseases based on nearly useless tests that arise from an outdated biological paradigm. Perhaps the most disturbing modern example of the abuse of testing technology comes from an earlier era of medicine when we witnessed the excessive use of the misleading PSA test for prostate cancer. (PSA stands for *prostate specific antigen*.) Sadly, *Scientific American* recently reported that "millions of men have gotten unnecessary biopsies, surgery, and radiation as a result of taking the PSA test."[10] Could such gross examples of malpractice based on genetic testing be in our future?

Ultimately, the usefulness of your medical test depends on the truthfulness of your scientific paradigm and the maturity of your research based on it. Research into the nature of the epigenome and the gut biome is progressing at an exponential rate, and the more advanced paradigm of systems biology that supports this research is leading to a new era of testing

---

10 John Horgan, "Why I Won't Get a PSA Test for Prostate Cancer," *Scientific American* (June 14, 2017).

that relies on the *biological assay*. The comprehensive gene tests of the past are now being *repurposed* within this broader "systems" context. Indeed, the rich data these newer tests supply us can be reframed to support health optimization—rather simply feeding today's obsession with the treatment of disease in isolation from its true causes!

These tests of the future will be far more targeted than the broad-brush DNA assays that are produced today. Gene tests will be linked with data from other biological measures and "crunched" with the help of advanced software. Most importantly, this sophisticated biological information will be framed within a holistic model of human health in which our daily health practices will occupy center stage.

The Search for Modifiable Genetic Biomarkers

You'll recall that certain genes are penetrant—they dictate irreversible gene expressions such as eye color, height, hair texture, or rare diseases. There is no "high court of appeal" when it comes to such deterministic genes; you can't change the gene expression that produces your hair color, no matter how rigorously you pursue a healthy lifestyle. Because researchers now know the locations of such genes, their "loci" (and that of other nonmodifiable DNA material) is set aside in the typical gene test. Instead, they target a smaller set of genes or DNA regions that matter to human health, throwing their resources at finding variants that code for disease with high probability. As we've seen, those ahead of the pack seek out epigenetic or biomic markers that point to opportunities for health optimization.

Assuming such technical feats can be accomplished efficiently, the next logical step is to get proactive and ask better questions. For example, how do we run tests to locate crucial factors that we can manage *before* serious disease sets in? What are the *modifiable* biomarkers that govern the major processes in the body that are the most critical for health? A vivid example is *C-reactive protein.* An excess of this biomarker in our blood is a solid indicator of general inflammation; an unknown genetic process in the liver produces it, but all agree that its presence is a useful indicator of this disease condition. A simple blood test gives doctors a reliable measure of this marker.

As a clinician, I believe that the best use of today's epigenetic research is to narrow down the markers that are the *most* predictive. Once we have a selection of these most useful biomarkers, we should do further research to reduce this list down to a critical few—therefore providing an inexpensive basis for a very different kind of biological assay than we have seen in the past. My colleagues and I believe that we will eventually reduce this key list of modifiable biomarkers down to approximately 30-40 total, or even fewer. These crucial biomarkers should be the focus of today's epigenetic research, since, by definition, they will point to those gene expressions—those epigenetic markers—that influence the most vital biochemical pathways in the body. As we will see in Chapter 3, at least seven such biological pathways determine which major diseases or states of health we will experience in our lives.

When gene expressions exert their influence on these important pathways, they leave behind biochemical tracers of their activity. But isolating and studying these traces has not proved easy. Locating them is like finding the tracks of an animal after a fresh snowfall. Identifying the diseases that "tracks" in our body might signify is like reading footsteps in the snow to determine which sort of animal has passed through, where it was headed, and when it was on this section of the trail. In the same way, biological tracks in our body give us a sense of the direction, speed, and intensity of the influence exerted in us by the function of biochemical pathways.

In addition, no one marker or "footprint" can ever tell the whole story. In fact, no single biomarker has decisive meaning all by itself. It is the *pattern* of "the footsteps on the trail" that must be interpreted by highly trained doctors, often working with genetic counselors.

Further, researchers are not in precise agreement as to which combination of markers point to specific health or illness outcomes. At a minimum, many varieties of data points will be needed to get there. I believe that these must include specialized genetic and epigenetic tests; the results of advanced blood tests; and the data from *biomic assays* (tests based on a sample from the gut microbiome)—along with other measures still to be determined. In theory at least, all of the biomarkers in each of these categories are modifiable. This mass of data can then be correlated using advanced software algorithms to produce a series of diagnoses and treatments consistent with the goals of personalized medicine.

It will not be easy to narrow things down to a small set of biomarkers and patterns among this vast range of data. According to one estimate, there are thousands of epigenetic markers; hundreds of indicators in a complete blood count test or CBC (a blood test used to evaluate your overall health); and hundreds of markers in a *biomic assay*. The permutations of these elements are in the millions!

Biological Tests of the Future

Today's exclusive focus on data-intensive gene testing will gradually become a thing of the past. The coming era of precision medicine will also give rise to *epigenetic* mapping—and we're fortunate that an international effort to profile complete human epigenomes is now under way. One of its first findings, for example, revealed that people with Alzheimer's disease had epigenetic changes related to their immune system, opening up a surprising new avenue of research. This clinically significant finding is just one of the many results of the work of the International Human Epigenome Consortium (IHEC), which was launched officially in 2010 in Washington, DC. IHEC says that it aims to produce over one thousand "reference epigenomes" and make them available to the international scientific community. Mastering this highly complex challenge will require another decade or two to be realized.

In the meantime, the hype regarding gene mapping will persist for a few more years. During the transitional period, government regulators are responding to the growing number of companies who market direct-to-consumer gene testing. As far back as 2012, the Government Accounting Office (GAO) purchased genetic tests for identical DNA samples from four of the most prominent companies in this field, and compared the test results for 15 common diseases. The reports varied greatly between the companies. A GAO spokesperson stated, "We found that 10 of the 15 companies engaged in . . . fraudulent, deceptive, or otherwise questionable marketing practices. . . . In general, [direct-to-consumer] testing is of little to no medical value."[11]

To further illustrate the "wild west" state of the genetics testing market, in 2013 the FDA ordered 23andMe—an early leader in the field—to suspend its operation. Its testing services assessed the risk for more than 250 diseases; consumers simply ordered a $99 kit directly from the company. Among their many worries, FDA regulators were concerned that false positives from the assay could cause some patients to receive excessive or unneeded medical care.[12] Thankfully, 23andMe has reformed itself and moved on, as we'll see in Chapter 6.

The next generation of biological tests is almost here. The

11 Associated Press, "Gene Mapping for Everyone? Study Says Not so Fast," http://www.dailyherald.com/article/20120402/business/704029816/ (Apr. 2, 2012).

12 Alberto Gutierrez, "Warning Letter, 23andMe, Inc.," Inspections, Compliance, Enforcement, and Criminal Investigations of the U.S. Food and Drug Administration, http://www.fda.gov/ICECI/EnforcementActions/WarningLetters/2013/ucm376296.htm (Nov. 22, 2013).

science of epigenetics is fostering a new regime of advanced tests for biomarkers across many biological systems, including the epigenome and biome. These test results will help you focus on how your life choices condition your gene expression. It is this element of choice that is lost amid today's hype about genome mapping.

Junk DNA: Shedding Light on the "Dark Genome"

To better understand how today's confusion among biomedical paradigms has come about, allow me to backtrack into what today seems like ancient times in genetic research.

To get started, let's return to the cozy scene of our DNA coiled up in the cell's nucleus in the form of 46 chromosomes. Next, let's isolate one of these chromosomes and magnify it under a very powerful microscope. What we'll discover is that thousands of genes are interspersed like Christmas-tree decorations along each DNA strand. The function of the genes found along this expanse, it was once thought, was to code for proteins that do the work of cell metabolism. By the year 2000, scientists had identified roughly 100,000 types of proteins in the human body; and because of their simplistic model of the genome, researchers expected to find about the same number of protein-coding genes. In other words, because of their mechanistic picture of human biology, scientists believed that there was always a direct relationship between a single gene and a single protein, and that the work of these proteins alone determines our health and longevity.

That's why they were truly startled to learn from the results of the Human Genome Project that the actual gene count is only about 20,000. Additional research went on to show that the protein-coding regions of human DNA (the locations where we find actual genes) account for *less than three percent* of the entire genome! Scientist were dumbfounded. Could the rest of our DNA really be useless junk, or remnants of the past that no longer contribute to cell metabolism? The puzzled genetics research community coined the term *dark genome*, a direct allusion to the concept of dark matter used by astrophysicists to designate the invisible and mysterious form of matter that makes up about 90 percent of the universe.

Significant technical progress began to point the way out of this conundrum. By about 2003, geneticists could trace all of the steps in the process of coding (or synthesizing) a protein. First, the code has to "go mobile," so to speak; special protein messengers have to be created from the raw genetic code that sits in a fixed position in the cell's nucleus. Serving this function is DNA's famous chemical cousin, *ribonucleic acid* (RNA), and scientists have become increasingly impressed in the years since by how versatile and clever this molecule can be. It's been long known that RNA literally "unzips" the coiled-up DNA strand, identifies a discrete portion of it, and attaches itself to this region in a kind of one-night stand. The RNA then "copies" the exposed gene instructions in that DNA region, which it treats as a template. The RNA zips back up the location it is working with, and then (typically) carries these orders outside the nucleus to the cell's "protein factories" that are located in the *cytoplasm* (the term for everything else in the cell other than the nucleus and the cell membrane). Here the RNA's copy of a portion of the gene code goes to work, specifying the amino-acid sequence required for a particular form of protein synthesis. These highly specialized proteins go on to perform any one of thousands of routine biological functions. And while the general outline of this picture was understood by scientists in the early days of the genetics revolution, better research tools have allowed them to fill in details down to the smallest molecular components.

Protein synthesis was now well understood, but what about the much longer patches of "junk DNA"—that mysterious portion of the genome that seems to sit there quietly filling in the space between (and even within) genes? It was known that these regions do not code for proteins, so how could they possibly be useful?

If all the kinks and folds of the DNA helix were stretched out instead of being curled up inside the cell's nucleus, this string would be about nine feet long. At this point, science knew what only *three inches* of this strand was doing when it interacted with RNA—a rather humbling situation considering that this embarrassing state of affairs was the case only about a decade ago. And further, what about the other sorts of RNA molecules that were now being identified, thousands of them in fact, that do *not* create proteins but seem to have other and unknown roles in cell physiology?

For better illustration, imagine that the unfurled helix was straightened out and magnified so that it extended the entire length of the Pacific Ocean, forming a bridge from California to Japan. At this level of magnification, our identifiable 20,000+ genes would look like thousands of dots of land—tiny islands that resemble stepping-stones across the great expanses of the ocean. Imagine tiny cargo ships hauling proteins out of the little ports on these islands. This archipelago of protein commerce would be like hundreds of little versions of the Hawaiian or Fijian islands, whereas the vast majority of the rest of the expanse would be very long tracts of water with a DNA double-helix bridge running over it. What was happening along this vast DNA bridge running over these open stretches of water to connect the bitty little islands of genes?

There's a big reason why this mystery eluded us for so long: cell metabolism is utterly complex and nonlinear. In other words, our metabolism is, as we noted in Chapter 1, the result of a vast network of diverse biochemical influences—a stark contrast with the image that is invoked by a machine-like correspondence of one gene with one protein.

More specifically, studies have now confirmed that the so-called dark genome contains vast DNA regions that *code for RNA*, doing so not to synthesize a protein but to create *other* types of RNA; indeed, many unexpected varieties of RNA were identified. And the plot gets even thicker. We've long known that RNA plays a big role in gene regulation and other cellular functions, but there are *also* long stretches of the DNA helix that do not encode RNA at all. These "empty spaces" in the Pacific Ocean actually regulate gene expression in yet *other* ways we are now learning about.

Along most of these stretches, and very often directly on and around the genes as well, chemical markers get imprinted in response to that "cloud" of multiple influences—that huge network of biochemical actors that are always impinging on the cell's environment. These influences carry out our epigenetic programming, utilizing a special epigenetic language for directly switching on or off, or up and down, the expression of specific genes or the other essential regions of our DNA that code for RNA or other molecules.

The ENCODE Research Project and Junk DNA

To better understand these frontier spaces in our DNA archipelago, big things have had to happen. "Big data" that was not derived from previous studies of actual genes had to be collected. This means that scientists had to penetrate the "darkness" of the *intergenetic* spaces in the DNA strand. And they finally did so in the form of a massive project called ENCODE ("Encyclopedia of DNA Elements").

This heroic endeavor was undertaken by an international consortium of 32 research institutes organized by the

International Human Epigenome Consortium, an umbrella organization we touched on in the last section. These scientists pooled their efforts to answer the long-standing mystery as to what was hidden in the 97 percent of the total human genome we knew almost nothing about.

Because most of the results of this project are in, we can cut to the chase. In September 2012, the study reported three major findings:

> **First, the "junk DNA" designation was starkly incorrect.** According to ENCODE's findings, "about 80 percent of the genome is biochemically active"—far beyond what was ever imagined to be the case.

> **Second, the researchers discovered that four million intergenic "spaces" on the DNA strand actually act as switches that control gene expression through RNA and by other means, and these switches are called *regulatory DNA*.** This finding also made it abundantly clear that the biochemical regulation of genes is more intricate than anyone ever expected. Because of this complexity, predicting specific diseases turns out to be more difficult than anticipated due to the staggering number of variables.

> **A third major insight from the ENCODE project is that disease usually occurs when a *structurally normal gene suffers from abnormal regulation*.** This means that

> searching for a single, abnormal gene is usually beside the point. It would be more expedient to research the vast stretches of the helix strand for receptor sites for specialized molecules that control gene expression.[13]

In short, most of what we once called "junk DNA" is actually a complex intergenic system that regulates genes; it's not the only epigenetic system hosted in the cell, but this once-mysterious domain of regulatory DNA is one of the several major types of epigenetic modifiers that help all living things adapt to their life conditions.

Further, this discovery has big implications for human health. The original ENCODE researchers found epigenetic switches spread across these regulatory DNA regions that are linked with cancer, multiple sclerosis, lupus, rheumatoid arthritis, Crohn's disease, and celiac disease, and their successors are going much further. Dr. Eric Lander, a leader in the Human Genome Project and now the president of a joint research endeavor of Harvard and MIT, has observed that the newly emerging understanding of intergenic DNA is a "stunning resource."

ENCODE results are also transforming cancer research. As the ENCODE team focused on cancer began determining the DNA sequences of cancer cells, they realized that most of the thousands of DNA mutations in cancer cells were not

13 E. Pennisi, "ENCODE: Project Writes Eulogy for Junk DNA," *Science* (Sept. 7, 2012): 337:1159. See also J. R. Ecker et al., "Genomics: ENCODE Explained," *Nature* (Sept. 6, 2012): 489:52.

occurring in the genes of these cells. *Instead, these mutations are to be found only in the epigenome.* The challenge now becomes figuring out which instances of epigenetic mutation actually drive a particular cancer's growth. Within this cancer team, one subgroup examined prostate cancer genes that, it was already known, are not readily attacked by drugs. They showed which regions of the epigenome control those genes, giving doctors an alternate way to go after them: by targeting the associated epigenetic switches.

Because the epigenome is stunningly complex, the ENCODE project was technically daunting. Its advances were only possible because of major advances in DNA sequencing and computational biology; these researchers generated 15 trillion bytes of raw data, and analyzing the data required the equivalent of more than 300 years of computer time.

ENCODE and related developments offer great hope for the future of mapping an individual's epigenome and tracking the influences that shape it. For example, at different points over a person's lifetime, we will be able to create a picture of that person's changing "epigenetic state." Or, epidemiologists will be able create "epigenetic maps" of groups of people in a specific locale to help explain their biological relationship with their immediate environment. The upshot is that our genes are swimming in an ocean of influences that determine their expression through a rich variety of epigenetic regulatory mechanisms. Mapping that complexity and finding clinical applications for this knowledge is the challenge before us.

A More Advanced Look: Gene Variants as Disease Biomarkers

A fly ball is going deep into the outfield, so let's return for a second look at an important fundamental: the problem of gene variants. We've noted that certain of these mutations, such as the gene for Huntington's disease, are inescapable; there's no stopping the expression of such rare genes short of physically "editing" them out of the DNA strand—a topic we cover in Chapter 6. Fortunately, almost all of these dangerous "one-trick" variants are known to science. Thanks to ENCODE and other efforts, research has now moved toward the wild-cat world of big data—a forbidding place in which computationally sophisticated scientists look for groups of genes or regions of the epigenome that are only indirectly associated with diseases. These studies are generally known as "gene associations." Such associations are useful factors in the equation that don't necessarily lead us to the causes of diseases, but instead can be correlated with more complex patterns or "clouds" that ultimately tell the whole biological story. These genetic or epigenetic associations can, however, act as *biomarkers* for those disease-creating patterns.

Targeting such general associations may sound less glamorous than discovering single-gene diseases, but this difficult research is still an important contributor to the era of personalized medicine. Such linkages can lead us to more customized strategies that are an incremental improvement on the current one-size-fits-all approach to much of medical care. In addition, if a patient does become ill, knowledge of gene associations can help doctors select the treatments most likely to be effective and least likely to cause adverse reactions.

The good news is that in some cases, these genome-wide association studies (GWAS), which we touched on before, have identified very specific markers that seem to have statistically significant connections with particular conditions.

The GWAS approach begins with the *reference human genome sequence* that was first produced by the Human Genome Project and has been refined ever since with the new data sets made available by ENCODE and other research projects. Using certain advanced tools, scientists search the whole genomes of thousands of people with particular diseases or traits in search of small gene variations that stand out when compared to the reference data. The scientific name of these variants is *single nucleotide polymorphisms* or SNPs (pronounced "snips"). Generally speaking, specific SNPs are found more frequently in people with a particular disease or trait than in people without these characteristics.

Here is a representative list from among the many exciting findings that have recently been reported by GWAS researchers around the world. These may seem "far afield," but they represent possible solutions to a great deal of human suffering.

1. Some SNP variants are located in proximity to sites in the genome that contain regulatory DNA. As a reminder, these DNA sequences are not genes; instead, they act as epigenetic regulators on the vast intergenic portions of the DNA strand. The teams discovered many disease associations to mutations (i.e., SNPs) found at specific regulatory locations.

2. A targeted GWAS has provided important clues about the genetic basis of age-related macular degeneration. Five major gene variants are now associated with this condition, and the presence of each is associated with up to triple the amount of risk for this eye disorder.

3. Another genome-wide association focused on a specific disease was able to identify more than thirty variants related to Crohn's disease. In particular, three of these SNPs are both very common and are directly associated with increased risk for Crohn's.

4. Important advances have been made in the genetics of autism because of a well-designed Chinese study. This effort was not at the scale of a GWAS, but it was equally ambitious. It found very strong evidence that the autistic children had mutations in at least three specific genes. This study represents progress, but the researchers pointed out that the work of sorting out the complexities of this mysterious disease is still in its infancy.[14]

---

14 In this study, three teams independently studied 549 families in which one child suffered from autism but the child's parents or siblings did not. Each team sequenced every gene in each individual (i.e., the autistic child, its parents, and some unaffected siblings). All three groups independently came up with the same basic finding.

5. GWAS researchers, also in China, have found strong gene associations to major depression in people who have endured excessive stress. Scientists previously knew that the degree of a person's stress and life adversity is linked to the incidence of two important genetic factors. Researchers conducted a GWAS called CONVERGE to look further into this association, and discovered that these same two biomarkers are also associated with the incidence of major depression in women.[15]

6. Several genome-wide association studies have identified genetic links between disease conditions that were previously thought to be unrelated. For example, an unlikely link between macular degeneration and inflammatory bowel disease was discovered in one GWAS. Other GWAS-inspired discoveries of diseases that share genes with ostensibly unrelated illnesses include type II diabetes, melanoma, Crohn's disease, Parkinson's disease, and prostate cancer. Finding these common pathways also underscores the potential for developing drugs or nutrients that may be effective in treating such formerly unrelated conditions.

---

15 The two genetic factors involved here were (1) the increased presence of mitochondrial DNA (energy-conversion structures outside the nucleus of a cell that host tiny amounts of DNA) and (2) shortened telomere "caps"—biochemical entities at each end of a chromosome that protect it from deterioration. CONVERGE was a GWAS of 5,864 women with recurrent depression who were compared to 5,783 women without depression histories. In this connection, it is interesting to note that the same telomere and mitochondrial changes can be reproduced by the administration of a naturally occurring stress hormone.

7. A GWAS's can help match an individual's overall genetic profile to the likely effect of particular drugs. In some instances, a person's genetic makeup can predict the occurrence of toxic side effects from a specific drug. Further, GWAS researchers have found that the presence of particular mutations in tumors can predict whether or not specific drugs will work or not as an effective treatment for that tumor.[16]

Nearly 600 genome-wide association studies covering 150 distinct diseases and traits have been published, and some observers have questioned their cost-effectiveness. But I believe that these findings reveal the increasing importance of computational biology to personalized medicine, both because of the uniqueness of each person's genome and also because meaningful biomarkers can now be discerned across large groups of people.

16 For example, researchers at Sloan-Kettering have found that it can be better to base cancer treatments on the genetic mutations generated by a cancer than on the organ system where that cancer originated. Conventional treatments might yield a response rate to chemotherapy of 10 or 20 percent, but the newer drug targeted according to the type of mutation has an amazing response rate of 50 or 60 percent. The National Cancer Institute is now doing an advanced research program along the same line.

Biomarkers and the Era of Personalized Medicine

Centuries of evolution in the medical profession have delivered us to the brink of truly personalized medicine. We now live in an era in which high-tech methods will soon yield cost-efficient ways to test for those biochemical markers that matter most to each individual's health. We'll interpret this test data by using for reference the results of thousands of studies of the impact of diet, drugs, stress, and exercise; the results of routine gene tests that indicate vulnerabilities; the new reference models of epigenomes and gut biomes coming out of universities; the results of genome-wide studies; and data from many other kinds of other advanced methodologies.

I have mentioned previously that—as our own contribution to precision medicine—my colleagues and I have been compiling a small number of what we believe are accurate, predictive, and inexpensive biomarkers. We began this process several years ago when we realized that no single laboratory test existed for all of the blood biomarkers we then believed should be tracked; to test for all of them, we realized we would need to send blood samples to five or six labs. So, several of my medical colleagues and I tried this awkward approach as an experiment. It cost us approximately $8,000 to test one of us, and the results were unimpressive. Every lab measured by a different standard, so a test of the same gene or biomarker at different labs inevitably yielded conflicting recommendations. Even the trained geneticists on our team found it difficult to piece together a cohesive interpretation.

This experience increased our determination to create a single laboratory that will focus on a small but meaningful set of markers, which I report on fully in Chapter 6. According to our protocol, the test result from that single source would be used to create a highly personalized plan that covers all the bases: diet, exercise, meditation, stress management, and other lifestyle recommendations. Then, at the end of 12 weeks, we would retest. Most of the markers should show improvement at this point. We would continue to make adjustments based on feedback received at regular intervals. Such personal fine-tuning at the level of genetic expression is unprecedented in human history. Will it give us radical new levels of health? Time will tell, but I am hopeful.

We can anticipate that the science will improve as we go. Right now, for example, there are five good markers for inflammation, but as we make progress, we may find that we only need to test for one of these. So, instead of testing for 30–40 markers overall, we may even find a way to reduce the list of markers to 15 and cut the cost even further.

As clinicians, we are also hopeful that patient compliance will improve as more evidence comes in with each new assay. But let's be honest. If you have to give up unfermented soy and a long list of other foods or habits you like "for the good of your health," the odds are you'll change only a few behaviors at first; then you can move on and improve your choices the next time as you begin to see useful results and feel better.

Our lifestyle choices may hold the key to our health, but habits are notoriously difficult to change.

We can also expect inevitable resistance from the conventional medical community itself. It takes years, sometimes generations, before any major innovation becomes widely accepted. Throughout history, scurvy (a vitamin C deficiency) has cost millions of lives. Yet its cure had been repeatedly discovered and forgotten for more than 400 years. Implementation of personalized medicine on a large scale will require a major reform of the healthcare infrastructure. Such a shift is bound to require years, if not decades. But such a transformation is a virtual necessity because of the profound implications of the epigenetics revolution and its related disciplines.

3

EPIGENESIS

Seven Pathways to Optimal Health

Because epigenesis guides the expression of our genes, the new science of epigenetics is fast becoming an essential guide, both to the treatment of diseases and in the quest for longevity and optimal health.

A good example is the true story of a 42-year-old patient we'll call Margaret, who had suffered from a series of more than 30 ear infections. Hoping to ward off the next attack, her doctor advised her to keep a supply of amoxicillin on hand. Margaret began to take this antibiotic every day, treating it almost like a daily vitamin supplement, but she would always get another ear infection. Meanwhile, her abuse of the amoxicillin wreaked havoc with her gut biome, killing beneficial bacteria and promoting an overgrowth of harmful bacteria. After enduring innumerable regimens of antibiotics without success, she turned to an alternative medical doctor as a last resort.

After listening closely to her list of symptoms, her new doctor diagnosed Margaret with irritable bowel syndrome and ordered several genetic tests. When her genetic tests came back, they revealed a genetic vulnerability to certain grain allergies. The only grain that she had been eating had been corn. Taking an epigenetic approach, her doctor recommended that she simply eliminate corn from her diet.

We've emphasized that epigenetics is the science of the alterations in gene expression that occur in response to changes in our environment or lifestyle. And this understanding provided an elegant solution for Margaret. When she stopped eating corn and corn products such as high-fructose corn syrup and the filler in her frozen dinners, she had chosen a new dietary lifestyle, thereby triggering an epigenetic effect. This single choice switched off the genes that produced her inherited allergy to corn. After ten days her earaches improved.

When the ear infections were gone, she stopped taking amoxicillin. After three weeks of living symptom-free, her doctor did a simple follow-up test—he asked Margaret to eat corn again. Within 20 minutes, she reported ear pain, and within an hour, fluid was extruding from her ears. She immediately returned to her corn-free diet. When she did, her symptoms resolved again without the antibiotics. Margaret's story is a vivid example of the epigenetic management of a genetic vulnerability that I will discuss at length in the next few chapters.

Unfortunately, the story is rarely as straightforward as it was in Margaret's case; you'll recall the earlier discussion about how our genome is bathed in a "biological cloud" of numerous influences that shape outcomes all at once.

To throw a different light on this point, let's think of the influences on our genome as an "ocean" of stimuli and regard each cell in our body as if it were a fish in a great sea. Because the ocean of biological influences on our cells is so vast, it is usually not possible to identify how a single input is going to act on a specific gene. Instead, it is preferable to think in terms of how each action we take might in some small way modify our relationship with the ocean as a whole. On rare occasions that impact is overwhelming and unmistakeable, such as what might happen to a fish in the Pacific Ocean if it was exposed to the poisonous radiation still pouring forth from the Fukushima nuclear reactors in Japan. In these severe cases, the genetic switches are overwhelmed and pushed in a negative, disease direction. However—and fortunately, most often—the influence from any given factor is much more subtle, making infinitesimal but sometimes vital changes in the ocean in which our cells are bathed. It is best that we remember this broad concept as we consider the specific impact of lifestyle choices on genetic expression.

But there is another major factor to consider. A fish or a cell doesn't simply slosh around all day in a chaotic world of influences. A great ocean, for example, displays distinct patterns in the deep that modify the activities of its various species of fish. Just as there are lanes in the ocean in which whales always seem to migrate or certain zones where particular schools of fish might congregate, we now know that certain well-traveled biochemical pathways in our bodies can influence our epigenome—seven of which I will discuss in this chapter. As noted in the Introduction, those "lanes" of influence that give us the most profound clues about both disease and vibrant health are *oxidative stress, inflammation, immunity, detoxification, lipid metabolism, mineral metabolism,* and *methylation.* In other words, particular aggregations of distinct patterns of stimulus can have consistent and discernible biochemical impacts on the body.

Some of these are easily recognizable and well understood, such as the pathway of inflammation. For example, exciting evidence suggests that psychological states such as trauma, depression, and even job stress can switch on an inflammatory response in certain genes; and if those switches remain active over time, the long-term impact can result in destructive and sometimes permanent biochemical changes throughout the entire body. Also, it has long been known that overindulging in certain forms of saturated fat turns on the expression of inflammation by the genes that regulate this response. This can lead to chronic inflammation in the arteries that in turn causes the buildup of cholesterol plaques, a major risk factor for heart disease. By contrast, studies show that eating the proper fats literally turns off the genetic signal for inflammation and leads to a state of regeneration and health.

These conclusions about the beneficial effects of good diet on inflammation carry a high probability of being true, but the impact of lifestyle choices on some of the other pathways are less obvious. What's more, many of our long-held beliefs about these matters may one day turn out to be wrong—the result of the daunting complexity of the influences on our genome from the "ocean" around our cells. Much more research is still needed on all fronts.

In a sense, we're like the boys in *The Maze Runners*, the recent science fiction movie. As we manage our health, we need to be clever enough to navigate the maze of the human body to find the right path to wellness. This is a very personal quest that each of us pursues with the help of our healthcare providers. The best approach to this challenge—in the face of so many unknowns and so much health data that is sometimes contradictory—is to find methods of testing and retesting our interventions. I believe that we need ways to assess, over time, the outcomes of lifestyle modifications and specific medical treatments through the comprehensive biological assays that I broached earlier, and which I explain more fully in the final chapter.

As we run this biological maze, I believe that we should rely on the following principles of personalized medicine that I have developed with my colleagues. This new era of personalized medicine is characterized by the "Four Ps":

1. **Personalized,** that is, it's based on tracking down and measuring the unique set of biological markers that have the most influence on an individual's health.

2. Predictable, meaning that the treatment protocol tests how particular biomarkers respond to a given set of health practices or interventions and then uses this feedack to point the way to even more customized treatments in the next round.

3. Preventive, which means that our approach shifts the focus from illness to the proactive pursuit of wellness—or even supernormal levels of well-being—through the rigorous application of a wide range of health practices.

4. Participatory, or intended to empower patients to "carry their own health portfolio" so that they make informed choices based on access to their medical records and to health information that was once available only to medical professionals.

The Seven Biochemical Pathways

Ultimately, our best approach to activating this list of principles is to monitor the seven biochemical pathways that govern our state of health, as well as to measure for the biomarkers that map into them. We do know a lot about the pathways; studies show that they underlie virtually all forms of disease; and we are discovering how these lanes of biological influence hold the keys to radiant health.

In this chapter, I introduce the pathways and explain

their roles. In the chapters that follow, I will detail specific changes in diet and stress reduction that correspond with these pathways and explain how to employ these choices to improve health outcomes.

1. Oxidative Stress

Oxygen is essential to life, but too much of it can be damaging. In fact, an excess of oxygen can be corrosive to all sorts of things, organic or inorganic.

When an exposed piece of apple turns brown, that's evidence of *oxidation*; similarly, long exposure to oxygen in the air causes iron to oxidize, or rust. In somewhat the same way, our cells can "rust" when exposed to excessive oxygen in the blood or tissues. Damaging oxidation results not only from poor diet but also when we are exposed to stress, toxins, and infections.

When we metabolize extra oxygen, electrons get chemically removed from atoms that are exposed to this process, leaving them with a negative charge. Oxygen is obviously necessary for life, but when unchecked oxygenation occurs in our body, this exposure is destructive unless *antioxidants* come to the rescue to neutralize the electrical charge. *Oxidative stress* happens when the quantity of electrically deficient molecules exceeds the presence of their nemesis—naturally produced antioxidants.

The damaging by-products of too much oxidation are called *free radicals*. In their search for missing electrons, free radicals can cause biological "riots." The destabilized molecules from which they pilfer electrons turn into scavengers hoping to steal an electron from some *other* molecule.

It is well known that excessive oxidative stress can cause gene mutation or cell death. Many chronic illnesses, such as cardiovascular disease, macular degeneration, and asthma, are connected to the ravages of oxidation. Skin wrinkling and many unhealthy conditions linked to aging are also associated with having too many free radicals.

Epigenetic approaches to reduce oxidation have proved highly effective. Solid evidence shows that two genes, PON1 and SOD2, turn on oxidative stress, but these genes can also turn it off. The "off" setting can be triggered when certain foods and specific nutrients are added to the diet, and also when we engage in stress management practices such as meditation. For example, research has demonstrated that a tomato-rich diet of 11 ounces of tomato juice each day as well as the frequent consumption of walnuts decrease oxidative stress damage because these foods turn off the PON1 and SOD2 genes. In addition, one of the most delicious, effective, and functional foods to reduce oxidative stress is blueberries. All foods with natural pigments are antioxidant, including red grapes, red cabbage, lemons, carrots, oranges, green tea, and green vegetables. Among the specific supplements that can reduce oxidative stress vitamins A, C, and E, CoQ10, alpha lipoic acid, folic acid, zinc, manganese, and beta carotene.

2. Inflammation

Chronic inflammation is a causal factor in a wide array of maladies, including cardiovascular disease, obesity, osteoporosis, cancer, inflammatory bowel disease, periodontal disease, rheumatoid arthritis, asthma, and allergies. Fortunately, this pathway is easily influenced through nutrition and stress management. For example, extensive research demonstrates conclusively that excessive consumption of red-meat protein is a major cause of prolonged and damaging inflammation; eliminating or reducing red meat from your diet can change that equation quickly. And there are many other dietary and behavioral causes of inflammation whose treatment is well understood.

Two kinds of inflammation can occur in the body: *acute* and *chronic*. Acute inflammation is a natural immune response that is necessary for overall health. Think back to a time you accidently cut your finger or picked up a splinter in your foot. A certain amount of redness, heat, swelling, and bleeding naturally occurs around the wound because you have created an acute need for the healing effects that are contributed by the natural inflammatory response to the injury. In this healthy process, white blood cells get released from dilated blood vessels and quickly surround any bacteria that are entering the wound. These white blood cells release chemicals whose function is to battle these potential invaders. Once this acute response is completed, a normal feedback mechanism stops the inflammation response and things return to normal.

In the instance of *chronic* inflammation, this natural feedback loop does not work. The body's inflammation response does not wind down like it normally would after an acute episode, but instead remains stuck in an "overzealous" state of activation. If unchecked, this process eventually contributes to the chronic diseases I listed above.

Further, the link between chronic inflammation and genetics is strong; diseases tied to this condition can be correlated with pre-existing genetic susceptibilities. Many genes have been identified that regulate inflammation through different pathways.

Providentially, there are a number of foods and nutrients that reduce or minimize excessive inflammation by turning off the genes that increase inflammatory activity or turning on those that restrain it. Among the anti-inflammatory foods to always include in your daily diet are green leafy veggies, beets, bok choy, broccoli, nuts and seeds, berries, garlic, and ginger; others include omega-3 fish oil, coconut oil, and monosaturated fats such as extra virgin olive oil, avocados, almonds, cashews—and, thankfully, popcorn. Nutrients in the *flavanoid* family, especially curcumin (an extract of turmeric), quercitin, and resveratrol in particular have a potent impact on turning down the genes that create excessive inflammation. Flavanoids are responsible for the vivid pigments found in many fruits and vegetables. These substances are part of the beneficial *polyphenol* group of compounds, and are found in apples, onions, cherries, some citrus fruits, leafy vegetables, raspberries, black tea, green tea, red wine, and red grapes—or that can be taken as supplements. Other helpful supplements include the antioxidants listed in the last section, and vitamin D. And be sure to always limit your exposures to pollutants in your food, water, and home supplies, which are dangerous sources of inflammation.

If we can identify people with a genetic vulnerability to those diseases related to chronic inflammation, we can also match them with anti-inflammatory pharmaceuticals. This was recently discovered to be the case with statin drugs that are widely used for treating cardiovascular risk and disease. A 2015 study reported in *Lancet* found that people with the highest genetic risk for developing coronary heart disease (CHD) appear to derive more benefit from statins than people with a lower genetic risk. To determine this risk, researchers identified and described the effects of statins on 27 genetic variants with known risk for coronary heart disease. They were able to demonstrate that the high-risk group had an astounding 72 percent increased susceptibility to CHD relative to the low-risk people. The anti-inflammatory statin drugs offer the greatest benefit to this high-risk group.[17]

This finding is an excellent example of an accurate, appropriate, and clinically meaningful application of genetic assessments for personalized care for heart disease. In this instance, the causal role of inflammation in CHD is well known, as is the anti-inflammatory impact of statin drugs.

---

17 J. L. Mega et al., "Genetic Risk, Coronary Heart Disease Events, and the Clinical Benefit of Statin Therapy," *Lancet* (June 6, 2015) 385(9984): 2264–2271.

This approach combines the identification of specific genes that are known to activate a specific biochemical pathway (in this case inflammation) with the use of an effective drug known to have a positive impact on that pathway.

Chronic inflammation is an intricate web of dysfunction caused by an interaction of multiple genes, diet, stress, and the environment. Research evidence continues to shed new light on which foods and practices are beneficial to quell the damage from chronic, long-term inflammation. But even if you don't yet know your genetic predispositions, the food and nutrient recommendations I have offered in this section and in the next chapter will go a long way toward reducing excessive inflammation in your body.

3. Immunity

Immunity is a vital function of the human body that allows our cells to distinguish "self" from that which is "not-self." Our body uses this special system to maintain its basic integrity and its overall identity.

Our immune system actively manages itself by generating what are known as B- and T-cells—true "memory cells." These cells are able to recall which disease-causing agents our bodies have previously encountered. When we come in contact with one of these "remembered" agents in the future, our immune response builds on the "lessons" of the previous encounter—when it learned, under duress and on the fly, how to remove an "alien" threat. That's why we call this function *adaptive immunity.*

What we eat and how we respond to stress has a profound impact on the immunity pathway, but it is important to remember that when it comes to our immune response, more is not always better. Instead, an optimal immune-system response is one that is balanced, appropriate, and moderate in its action. An overly aggressive immune response can lead to the development of *autoimmune disorders* such as arthritis, inflammatory bowel disease, and MS. At the other extreme, an under-response to an external threat leaves us susceptible to infection.

Quite a few Americans suffer from the hyperactive immune conditions such as allergies, asthma, and the many autoimmune diseases, and genes play a big part in all these maladies. For example, if one of your parents has an allergy, you have a 25 to 50 percent chance of getting one; if both parents have allergies, your chances can go as high as 75 percent. Research with animals shows that allergies and autoimmune illnesses are linked to the same genes.

Chapter 4 on diet and Chapter 5 on stress management provide you with practices that will have a profound and positive impact on this pathway.

4. Detoxification

Detoxification is a critical function in every cell and pathway. Detoxing is crucial, first of all, because the natural functioning of cellular metabolism produces toxic by-products. These waste products need to be removed to maintain our health, much like fallen debris must be moved from city streets in order to maintain traffic flow; similarly, when organic debris accumulates in our bodies, the toxic buildup interferes with the ordinary "traffic" of our cells. Because healers in every culture have recognized this issue, regimens of detoxification have been used all over the world for hundreds of years. But the heavy load of toxins to which we are now exposed makes a regular practice of detoxification an almost urgent need for all of us. For that reason, I offer specific instructions in this section to guide those who are new to this practice.

It is difficult enough to eliminate natural waste materials; but these days our cells must deal with an unprecedented overload of toxins from our air, food, and water that they also must eliminate. Consider the fact that the average American consumes literally pounds of hormones, antibiotics, food chemicals, additives, and artificial sweeteners each year, and that each one of these toxic chemicals has been shown to harm the brain. In addition, we consume about a gallon of neurotoxic pesticides and herbicides each year just by eating conventionally grown fruits and vegetables. (And that occurs even with people eating much less than the 8 to 10 daily servings of fruit and veggies that they should be eating!) Remember, pesticides are used by farmers because they are neurotoxic to pests—they attack their nervous system. Imagine what these poisons do to *your* nervous system!

What seems most unfair is that we expose our own children to obvious toxins. For example, a recent study in the *Lancet* clearly showed that food additives make children hyperactive. Researchers took about 300 normal children and split them into two groups. Each group was given an identical-looking colored drink. One was naturally colored, and the other contained sodium benzoate and many other colors and additives that are typically used in beverages for kids. The children who drank the tainted beverage were all much more hyperactive.

The neurological effect of the toxins in our food, water, and air is alarming. These compounds create changes in mood and trigger aggressive behavior, depression, problems with attention and focus, sleep problems, reduced intellectual performance, and memory loss. We don't get these results from just occasionally consuming one red candy, a blue cupcake, or the few drops of pesticide sprayed on our strawberries. It is the consistent, repetitive, cumulative presence of these chemicals in our lives that leads to broken brains and the epidemics of Alzheimer's disease, depression, and dementia we see in today's world.

We experience two phases of normal detoxification. In phase one, the liver uses enzymes to regulate oxygen, which in turn converts soluble toxins to more digestible forms. Phase two involves the conversion of these digestible by-products into water-soluble particles that can be excreted from the body. Both forms of natural detoxification can be directly enhanced and made more effective by eating specific foods. The scientific data indicates that both phases of detoxification are moved in a beneficial direction when diets contain *cruciferous vegetables* such as broccoli and cabbage, and *allium vegetables* such as garlic, onions, scallions, and asparagus. Other useful foods for cleanses are grapes, berries, green and black tea, as well as many common herbs and spices.

Here is additional advice to keep in mind as you pursue your chosen regimen of detoxification:

The first requirement is that you simply slow down. Make sure you include time for deep relaxation into your plan for a detox. Fatigue is normal during a cleanse, so allow more time for rest and sleep. Also, meditation, deep breathing, or any calming activity is good during a detox. Actively engaging your parasympathetic nervous system helps restore your energy, which your body needs to replenish itself.

Second, bear in mind that unsavory side effects appear in the initial phase of a cleanse. Keep a watch out and even keep track of them in a journal. For example, constipation can occur, so be sure you remember to move those bowels. Bear in mind that toxins in our digestive tract may make us feel ill if we don't eliminate them. Taking 300 mg of magnesium citrate at dinner will help to avoid constipation and ensure that your bowel movements remain normal to remove all toxins from your body.

Other typical symptoms of a major detox include bad breath, achy or flu-like feelings, fatigue, hunger, irritability, itchy skin, nausea, offensive body odor, and sleep difficulties ranging from too much to too little. But don't worry—these are just temporary and normal expressions of the fact that your body is working to eliminate toxins. Facilitate the process by drinking plenty of purified water, and try to start your day by drinking warm water with lemon. Taking an Epsom salt bath can also be helpful.

A second major side effect to keep track of is food allergies and food sensitivities that can erupt once they are no longer masked by the toxic overload. As your body increasingly lets go of toxic waste, it will be easier for you to recognize a hidden reaction to gluten, dairy, soy, or any of the other common food allergies. In more severe cases, eliminating food allergies and unjunking the diet causes symptoms similar to withdrawal from addictive substances like caffeine, alcohol, nicotine, cocaine, or heroin. Getting off of these allergens can cause a brief, flu-like achy syndrome that may last one to thee days. Keep in mind in this connection the odd fact that we are often most addicted to the foods we are allergic to.

Initially, those who consume the most caffeine, alcohol, and sugar, and those who have the most food allergies, will have the most difficulty. It is best to slowly reduce your intake of caffeine, alcohol, sugar, white flour, and over-the-counter medications (except as directed otherwise by your physician) a week or two before you start your program. Eliminate all refined sugars, flours, caffeine, alcohol, dairy, gluten, or addictive substances. If you allow these triggers to stay in your diet, the body will keep you tied to a vicious cycle of cravings and addictive behavior. Instead, you want to reset your biology to eliminate all such triggers. During the fast, drink at least six to eight glasses of filtered water daily. Stay away from plastic bottles and use glass bottles.

To boost energy during your cleanse, exercise for 30 minutes a day. Walking outside in fresh air is best. Roll up those sleeves and let the sun hit you with some vitamin D! Also, take 1,000 mg of buffered vitamin C with breakfast and dinner.

When on a detox, don't wait until you are starving to eat! Balance your blood sugar by eating protein-based meals and snacks every three to four hours. Excellent sources of protein are baked or broiled fish, lean poultry, and legumes such as edamame or black beans.

Heat is a great resource while cleansing, as it helps draw out toxins from your intestinal tract and entire body. You might also try a sauna or a warm bath with Epsom salts for 20 minutes a few times per week, since your skin is the body's largest organ of excretion.

If you follow these guidelines, you should feel better in three to seven days. But if you do not feel well at this point, please exercise caution and check in with your doctor.

Finally, please note that many of the dietary practices discussed in Chapter 4 can also activate this pathway and improve your efforts at detoxification.

5. Lipid Metabolism

Lipid metabolism is the fancy medical phrase for how our body breaks down, stores, and utilizes dietary fats. Because we've been hearing for years about the connection between excessive fat consumption and coronary heart disease, too many of us have drawn the conclusion that fats in general are unhealthy for us. In fact, the intake of healthy fats in our daily diet is essential for the creation of hormones, enzymes, and other vital elements of our body chemistry. What is most important is not the amount of fat in your diet, but which fats you are eating and whether these healthy fats are being used to good advantage.

The issue of fat consumption is critical in America because of our raging obesity epidemic. According to the Centers for Disease Control, 32 states have obesity in 25 percent of their population. In six of those states, the prevalence of obesity is at greater than or equal to 30 percent!

Obesity is evidence of a dysfunction in lipid metabolism, the root cause of which is not only diet but also a wide array of genetic influences. Over 50 genes have now been linked with various forms of poor lipid metabolism, and some genetic tests are now available to help determine your particular form of lipid dysfunction if you suffer from obesity.

Even without such testing, there are certain dietary guidelines that have a beneficial epigenetic impact on virtually all the genes responsible for obesity. My top recommendations include the following:

1. Reduce your intake of all forms of the "bad fats," which include *saturated fats* (such as beef, butter, and cheese) and *trans fats* (which are found in foods such as margarine or cookies).

2. Increase your consumption of monosaturated fats (such as nuts, avocados, or olive oil) listed earlier.

3. Decrease your intake of harmful carbohydrates (i.e., all refined flours).

4. Increase your consumption of high-quality carbohydrates such as whole grains.

5. Consume more cruciferous vegetables such as broccoli and cauliflower, which by the way are also able to switch on anticancer genes.

6. Increase the amount of folic acid in your diet by eating leafy vegetables, sunflower seeds, baker's yeast, and liver.

7. Increase your vitamin B12 intake by eating organic meats, liver, shellfish, and milk.

8. Obtain more vitamin B6 by consuming whole grain products, vegetables, nuts, and certain meats.

Overall, these recommendations are consistent with adherence to the Mediterranean diet, which I explain in detail in the next chapter.

Newer research suggests that we can change the expression of the underlying genetics governing our fat metabolism. Over the last ten years, researchers at the USDA Human Nutrition Research Center on Aging at Tufts University have observed epigenetic changes in the lipid metabolism pathway in response to dietary changes. In one study conducted in 2011, the researchers examined potential epigenetic markers of weight loss by studying twenty-five overweight and obese men who participated in an eight-week diet. The scientists compared the methylation patterns of men who had significant responses—both good and bad—to a calorie-restricted diet in terms of their weight loss. (See the section just below on methylation for details on this pathway.) At the end of the study, DNA methylation levels were greater in the good versus the bad responders to the low-calorie diet; vital sites on the DNA of the good responders were found to be altered epigenetically. Based on these findings, the researchers concluded that a calorie-restricted diet could alter the DNA methylation of specific genes governing lipid metabolism within an eight-week period. Moreover, they determined that the presence of this epigenetic marker could help predict which individuals would be most successful at weight loss.

6. Mineral Metabolism

Minerals are, of course, an important component of our bones and teeth. But they are also cofactors for energy metabolism, for creating and utilizing proteins, and for building red blood cells. Adequate dietary minerals are crucial in overall optimal health and longevity. In fact, individuals with chronic diseases often have disorders of mineral metabolism. In recent years, researchers have discovered epigenetic markers that can help us identify how well our bodies are metabolizing minerals.

Some disorders of mineral metabolism are linked to genes, but other diseases in this pathway result from either excessive or insufficient intake of minerals. Bear in mind that many of the commercially produced foods available in our supermarkets have only marginal mineral content.

At the present time, the two most studied minerals are salt and iron. Let's briefly consider the case of salt.

Salt has long been known to be an essential mineral that maintains fluid volume in the body. Our *renin-angiotensin system* (RAS) manages our blood pressure regulation by modulating the balance of salt and water in the body. RAS also manages our vascular tone—the degree of constriction taking place in our blood vessels. Through these two mechanisms, the RAS is known to influence blood pressure responsiveness in relation to our salt intake.

We have all heard of the "salt versus no salt" debate, but this controversy is really meaningless unless you are one of those genetically predisposed salt-sensitive individuals for whom excessive salt leads to dangerously elevated blood pressure. And this is no small matter, since 50 percent of us are genetically hypertensive, salt-sensitive individuals! The typical American diet contains excessive amounts of salt, which causes great harm to these folks.

Geneticists have identified two genes that produce the enzymes that control the RAS system. If you are hypertensive, a genetic assay can tell you if you have one or both of the genes that predispose you to high blood pressure. This simple test resolves the abstract "salt versus no salt" debate in a way that is personalized for your own unique metabolism.

7. Methylation

As we have noted, our daily experiences bring about all sorts of epigenetic modifications to our DNA. Very often, these subtle switches influence the pathways discussed above; but in addition, these biochemical regulators of the genome can themselves be considered to be among the seven biological pathways.

Every instance of epigenesis leaves a tell-tale sign of its presence known as epigenetic markers—and the best understood among these markers are the epigenetic molecules laid down by *DNA methylation*. In this process, a simple molecule know as a *methyl group* (CH3) controls certain aspects of gene functioning by attaching wherever it can find a C-base next to a G-base in a nucleotide, and there are innumerable such locations in the genome. The way a methyl group attaches to DNA has been compared to the way a sucker fish attaches to the belly of a shark; enough of them on the belly of a shark can change a shark's (or your DNA's) behavior. Research is now at the point where scientists can detect both the presence and the number of methyl markers in a particular location on the DNA strand.

Here's just one example of how they work. Abnormal methylation patterns are thought to be one of the root causes of excess cell proliferation leading to cancer. Two related biochemical processes are at work in such cancers: Tumor suppression genes can be overmethylated and turned off inappropriately, while tumor-promotion genes are undermethylated and are turned on incessantly.

More recently, advanced research has clarified other means of epigenetic modification—most notably types of proteins known as *transcription factors,* which control the transcription of genetic information from DNA to RNA. The resulting RNA, known as *micro-RNA,* is able to bind itself to a specific DNA sequence, thereby acting as its regulator (see also Chapter 2). Yet another important form of epigenetic regulation is known as *histone modification.*

In the near future, epigenetic maps which link the presence of methylation (or other types of epigenetic markers) to a particular condition will be an essential factor in controlling disease or turning on optimal health.

Are Epigenetic Markers Inherited?

It's a proven fact: In addition to inheriting DNA from our parents, we can also inherit epigenetic instructions on this DNA that comes to us "hard-wired" into our genome. While it is sometimes known as *soft inheritance,* these epigenetic modifiers can endure for many generations.

One of the easiest ways to demonstrate this paradigm-busting discovery is with studies of pregnant females and their progeny. Epidemiological studies as well as animal experiments have shown that maternal diet during pregnancy can produce epigenetic changes through altered methylation in the mother that are inherited by the offspring. For example, a famous 2003 study found that changes in the diets of pregnant mice permanently modified the methylation of certain DNA regions, leading to changes in the coat colors of their offspring.

To test whether a similar effect could be seen in humans, researchers studied 167 pregnant women living in a rural region of the Gambia. The diets of these women during their pregnancies varied greatly because of the effect of seasonal climate changes on their food supply. The scientists created a baseline by collecting year-round data from nonpregnant women. Through the collection of blood samples, the researchers found that the 84 women who conceived during the rainy season, when food was more plentiful, had significantly higher levels of nutrients in their blood in early pregnancy than the 83 women who had conceived during the dry season. Next, they tested six different genes in each child that these two groups of women gave birth to, looking for the presence of methyl groups on these genes. The researchers found that the genomes of children conceived during the rainy season had significantly higher methylation rates than children conceived during the dry season. Higher methylation rates were linked to higher concentrations of certain nutrients in the mother's blood, especially homocysteine and cysteine. It comes as no surprise that a poor diet led to lower methylation rates, as normal methylation requires the presence of nutrients such as choline, folate, methionine, and vitamins B2, B6, and B12.

In addition to diet, other studies have shown that a mother's exposure to toxins may also produce epigenetic changes in her offspring. According to Dr. Linda S. Birnbaum, director of the National Institute of Environmental Health Sciences and of the National Toxicology Program, "The susceptibility persists long after the exposure is gone, even decades later. Glands, organs, and systems can be permanently altered."

• • •

The importance of the seven pathways is clear enough, but scientists need to conduct much more research to identify those changes in the pathways that are likely to be causal in determining illness, optimal health, and longevity in relation to our genome, epigenome, and gut biome. Future studies should focus on the following:

1. Identifying epigenetic biomarkers that can be validated as indicators for specific conditions

2. Finding better ways to measure impacts by testing easily accessible factors such as saliva, blood, or stools

3. Refining our knowledge of lifestyle or medical interventions for which there is evidence of a positive outcome.

The seven biochemical pathways govern the foundations of health but can also act as signposts to disease conditions of all kinds. By adhering to the guidelines and recommendations in Chapters 4 and 5, we can influence our pathways in a positive direction. Once we engage in these practices, we will literally feel more energy and greater mental acuity. Above all else, we will know through our own experience that we can choose our own future states of health and longevity.

4

NUTRIGENOMICS

Your Genes and Your Biome Express What You Eat

Scientific evidence continues to pour in, telling us that our lifestyle and our environment—and especially our diet—are crucial factors in creating optimal health through improved gene expression. Because of these advances, we now know much more about how our daily intake of nutrients changes the biochemistry of the seven pathways, often by activating gene switches or by creating alterations in our microbiome. New studies even suggest that a meal can measurably change the gut's microbial colonies in *a single day*, for better or for worse, and that the biome governs certain epigenetic switches all on its own. This chapter offers my selection of the most important dietary studies of our time to help guide you in optimizing your epigenome and gut biome.

A landmark example of the centrality of nutrition is a massive long-term study of 23,000 people called EPIC, which was reported in 2016 in the *Archives of Internal Medicine.* Researchers wanted to know what, if any, health benefits would result from four simple behavioral choices: not smoking; exercising 3.5 hours per week; eating a diet of fruits, vegetables, beans, whole grains, nuts, seeds, and low red meat; and, maintaining a healthy weight, which they defined as a body mass index (BMI) of less than 30.

When researchers followed up almost eight years later, they found that the people who had adhered to those lifestyle changes had a 78 percent lower overall risk of developing some of the most devastating chronic diseases. Broken down by specific diseases, their findings revealed a 93 percent decrease in diabetes—a very impressive improvement. In addition, the researchers measured an 81 percent reduction in heart attacks; 50 percent fewer strokes; and 36 percent fewer cancers.

This remarkable result points us in the right direction, because the regimen of the study required both a healthy diet and a decent amount of exercise—on average 30 minutes per day. As I have argued for decades, we can weave an enduring tapestry of optimal health by making such wise daily choices based on the best science available.

But many confusing and complex questions remain to be solved by nutrition scientists. We're dealing here with a field of study that has been in great flux; in recent years, new research has turned the concept of best dietary practices on its head several times. Cholesterol, once definitely considered bad, can now be either good or bad. Carbohydrates, at one time the centerpiece of popular low-fat diets, are now implicated in the increased prevalence of obesity and diabetes. Certain fats and oils, once considered among the greatest of food evils, have demonstrated significant cardiovascular benefits among other beneficial effects—although red meat and other saturated-fat meats and trans fat oils still remain taboo. Such findings have led many to adopt low carb/high fat/high protein approaches such as the Paleo diet.

But then a huge study of 140,000 people in Sweden turned the tables again when its findings seemed to rehabilitate a diet featuring healthy carbs and low fat. "The results were categorical," stated the Swedish researchers. "[Our] study concludes that, over time, reducing animal fat intake *decreased* blood cholesterol levels, and that a high-fat/low carbohydrate diet *increased* blood cholesterol levels."[18]

This chapter encourages you to stay abreast of the best available science so you can make better choices for your unique genome that show up in positive biomarkers. I'll also show you how proper diet can eliminate major epidemics such as metabolic syndrome and other chronic diseases that are directly fed by poor dietary practices.

---

18 Ellen Shell, "Time to Retire the Low-Carb Diet Fad," https://www.theatlantic.com/health/archive/2012/06/time-to-retire-the-low-carb-diet-fad/258343/ (June 11, 2012).

The New Science of Nutrigenomics

Allow me to put it succinctly: certain vital nutrients literally *talk to our genes.* Advanced studies of the relationship between our genes and the foods we eat have created a powerful emerging science called *nutrigenomics.* Because this new "hybrid" science links our genes to our diet, we will soon be able to move beyond the questionable science behind today's popular dieting programs. On this basis, we'll be able to study the effect of nutrients on relevant biomarkers—almost in real time—by taking frequent assays of your blood, genome, and biome indicators and using this personal data to modify your diet so that it talks in the most optimal way with your genes.

Our first step in this inquiry is to look in on the commercial diets that bank on our desire to lose weight and look good. Unfortunately, several major studies reveal that common commercial diets show little evidence of sustained weight loss. Again, this finding contrasts sharply with what awaits us in a few years when personalized medicine will take the place of such hit-or-miss dietary practices. Let's survey the empirical evidence about popular dieting programs.

In a systematic review, researchers evaluated 45 studies (including 39 randomized trials of at least 12 weeks duration) in which commercial weight-loss programs were compared with a control group.

According to the results of this meta-analysis, the documented weight loss after 12 months from commercial programs relative to the control group was as follows: 4.9 percent weight loss with Jenny Craig; 2.6 percent loss with Weight Watchers; and for Atkins it varied between one percent and 2.9 percent. Nutrisystem and other very low calorie programs (including HMR, Medifast, OPTIFAST) resulted in short-term loss at three to six months but provided no longer-term weight loss. Generally, it was found that highly structured programs with the most in-person social support (notably Jenny Craig and Weight Watchers) had the best results over time.

As you can see, when it comes to weight-loss programs, the results are thin and the evidence of benefit is slim. Some weight-loss programs may be slightly better than others, but in the long run none has been able to produce significant or sustained weight loss, concludes the *Annals of Internal Medicine* review. For even the best programs, an editorialist writes, "weight loss is modest and likely below patients' expectations."[19] These findings complement those of a meta-analysis in the *New England Journal of Medicine* and yet another in the *Journal of the American Medical Association* in which researchers found that low-fat and low-carbohydrate diets yielded similar outcomes. The upshot is that while the amount of money Americans spend on such programs is staggering, the average weight loss resulting from this huge investment is modest at best.

Controlling our weight and eating right is, of course, an essential health practice. But how do we know what foods

19 Larry Husten, "Weight Loss Programs: Slim Evidence and Thin Results," https://www.jwatch.org/fw110061/2015/04/07/weight-loss-programs-slim-evidence-and-thin-results (Apr. 7, 2015).

to choose? Or, as my friend Dr. Mark Hyman puts it in the title of his insightful new book, *Food: What the Heck Should I Eat*?

First of all, our bodies, if they've not been abused by diet fads or personal food compulsions, have a built-in native intelligence—like our animal cousins—that naturally leads us to consume healthy foods; we knew this to be the case even before hundreds of diet books hit the shelves. In addition, the *empirical* evidence that supports beneficial dietary practices is also becoming preponderant. But even more important, the new science of nutrigenomics will soon be able to provide answers at a level of detail we've never had before.

Today's favorite diets range from "Paleo" or high-meat protein diets to the many low-fat and no-meat approaches and on to the Mediterranean diet. You are probably well aware that some of these well-established diets contain diametrically opposed recommendations. Yet all claim to have near-miraculous properties for losing weight and curing diseases. All have devotees who offer before and after photos with inspiring stories as evidence of their diet's unfounded claims. But how can anyone know what to eat to optimize their health, especially when the answer is different for each individual because of our differing genetic profiles?

The Mediterranean Diet—One of the Best

To date, the Mediterranean diet, one of the most evidence-based of all diets, is pointing us toward research directions that may provide answers to that question.

This approach consists of lots of vegetables, fruits, legumes, whole grains, herbs, spices, nuts, seeds, fish, and seafood, and recommends moderate intake of poultry, eggs, olive oil, cheese, and yogurt. It does allow for occasional meals with organically raised red meat or some processed meat and saturated fats, but the general thrust is to replace red meat with fish and reduce saturated fat intake by replacing butter with olive oil. Although it is difficult to discern which element of the Mediterranean diet is most important, it has long been regarded to be a prototypical example of a healthy diet and has been the subject of numerous studies suggesting benefits in regard to aging, cardiovascular risk factors, mood, cognition, and longevity.

A recent article published in the *New England Journal of Medicine* reported on a meta-analysis of 22 studies examining the effect of the Mediterranean diet on stroke, cognitive impairment, and depression, and found that high adherence to the diet was associated with a protective effect against all three.[20] However, researchers have been unable to determine whether this positive result is related to the well-known

---

20 Ramón Estruch, MD, and Emilio Ross, MD, "Primary Prevention of Cardiovascular Disease with a Mediterranean Diet," http://www.jwatch.org/na32643/2013/11/04/mediterranean-diet-multiple-benefits (Apr. 4, 2013).

anti-inflammatory or antioxidant properties of this diet or to some other cause, such as glucose regulation. Although we cannot yet verify in detail the specific beneficial components of the Mediterranean diet, its general principles are valuable and practical for lifestyle guidance for those who are not strict vegans or vegetarians.

The Mediterranean diet is especially thought to confer cardiovascular benefit. Recently, a team of Spanish researchers tested this belief in a trial of about 7,500 people ranging in age from 55 to 80 with no known incidence of cardiovascular disease. The scientists tested two slightly different Mediterranean diets against a control group on a low-fat/high-carb diet. After five years, this scientifically rigorous study made headlines when it was discovered that cardiovascular-related death occurred considerably less often in the two Mediterranean-diet groups than in the control group.[21]

Spain has become an epicenter of scientific interest in this diet. Another large Spanish trial assessed the efficacy of Mediterranean diets for the primary prevention of diabetes. Based on this seven-year study, the research team concluded that a Mediterranean diet, especially when supplemented with extra-virgin olive oil, reduced the incidence of diabetes among persons with high cardiovascular risk.[22]

---

21 Allan S. Brett, MD, "Mediterranean Diet Lowers Rates of Adverse Cardiovascular Events," http://www.jwatch.org/jw201303120000001/2013/03/12/mediterranean-diet-lowers-rates-adverse (Mar. 12, 2013).

22 J. Salas-Salvadó, M. Bulló et al., "Prevention of Diabetes with Mediterranean Diets," https://www.ncbi.nlm.nih.gov/pubmed/24573661 (Jan. 7, 2014).

Yet another big, long-term study of this diet (this time not in Spain) had eye-opening results. This trial focused on the telomeres of women who adhered to the standard Mediterranean diet. The outcome, in brief, was that these women retained significantly longer telomeres. The authors of the study calculated that these women's greater telomere length would result in an average gain of about 4.5 years of life! This means that the longevity impact of eating the Mediterranean diet is equivalent to the difference in life expectancy between nonsmokers and smokers, or between highly active and less active people. As a reminder, telomeres are repetitive DNA sequences located at the tips of chromosomes that protect chromosome integrity. Telomere length is a biological marker of aging; longer telomeres are clearly associated with greater life expectancy and lower risk for chronic disease.[23]

This diet's impact on actual longevity is a unique and encouraging finding. It is notable in this connection that a large Italian study (unrelated to diet) showed that people with shorter telomeres had a two- to threefold higher risk for developing cancer and that shorter length was also associated with a higher cancer fatality rate.

---

23 Paul S. Mueller, MD, "Mediterranean Diet Is Associated with Telomere Length, a Biological Marker of Aging," http://www.jwatch.org/na36616/2014/12/23/mediterranean-diet-associated-with-telomere-length (Dec. 2, 2014).

Strong Evidence for Vegetarian and Pesco-Vegetarian Diets

Well-designed large studies show that vegetarian or near-vegeterian diets reduce the risk of colorectal cancer and have many other important benefits. The findings about colorectal cancer are especially striking because this disease is the leading cause of cancer mortality.

The best scientific insight into this dietary approach dates back to the Adventist Health Study (AHS), which concluded in 1988. It found that Seventh-Day Adventists, a large Christian sect that practices vegetarianism, have a reduced risk of obesity, hypertension, diabetes, and mortality. This landmark finding has been cited by committed vegans and vegetarians ever since.

A systematic look at this group of committed vegetarians, known as Adventist Health Study 2 (AHS-2), was initiated in 2002 and concluded in 2007. In this study, a large cohort of 96,354 Seventh-Day Adventist men and women was categorized into four types of vegetarian diets followed by Adventists: vegan; lacto-ovo vegetarian (consuming eggs and dairy); pesco-vegetarian (consuming fish); and semivegetarian. The control group followed a nonvegetarian diet. This newer study showed results similar to AHS across all forms of chronic disease.

A relationship of vegetarianism to colorectal cancer risk had not been well-established until a supplemental analysis of the AHS-2 data was undertaken with this exclusive focus.

During many years of follow-up, researchers documented 490 cases of colorectal cancer. Compared with nonvegetarians, all the Adventist vegetarians combined had a significantly reduced risk for colorectal cancer. But it is interesting to note that, when the numbers were broken out by type of vegetarian diet, the pesco-vegetarians had the most significant reduction in risk. This fact seems to support the recommendation in the Mediterranean diet to engage in fish consumption.[24]

A Healthy Diet Supports a Healthy Brain

Because we now inhabit the era of mind-body medicine, nutrigenomics is also vitally concerned with the issue of how closely our daily diet is linked to our mental and cognitive health. This question is relevant for all ages, but especially for those who are elderly.

Two studies reported in *Neurology* in 2015 are among many in recent years that provide proof that good diets can especially support healthy cognition.

The first of these, which is based on the latest analysis of the data acquired in the well-known Framingham Heart Study, found that diabetes was a risk factor for both cognitive impairment and dementia. Researchers discovered that patients with poor glycemic control (a major indicator for diabetes) at the

---

24 M. J. Orlich et al., "Vegetarian Dietary Patterns and the Risk of Colorectal Cancers," https://www.ncbi.nlm.nih.gov/pubmed/25751512 (May 2015).

age of 40 went on to develop brain problems known as *gray matter atrophy* and *reduction in white matter integrity.* This finding lends support to other recent studies that also show a strong link between cognitive decline and type 2 diabetes.

The second *Neurology* study revealed that a healthy diet could reduce the risk of cognitive decline in a population over age 55 with diabetes or a history of stroke, heart disease, or peripheral artery disease. This research utilized the *Alternative Healthy Eating Index*, which emphasizes high amounts of fruits, vegetables, nuts, soy protein, a high ratio of fish to meats and eggs, and no more than a moderate consumption of alcohol.

In 2015, *JAMA Internal Medicine* reported that the Mediterranean diet is not only heart healthy but also brain healthy. For this study, 334 older adults were randomized into three groups: (1) Mediterranean diet plus extra virgin olive oil (one liter/week); (2) Mediterranean diet plus mixed nuts (30 grams/day); or, (3) controls who only received advice to reduce dietary fat. On a variety of cognitive tests, the Mediterranean diet groups scored better than the controls; interestingly, the olive oil group displayed improved memory testing by the conclusion of the study, while the nuts group (no pun intended!) had improved scores on tests of executive functioning. By contrast, the low-fat group showed declines in many cognitive domains.

Such results are significant for all of us who are aging. But what's most needed now is further comparison of such results with other studies around the world, as well as new studies of diverse populations over even longer periods of time, plus meta-analysis of all the data accumulated. In addition, scientists steeped in nutrigenomics will need to add epigenetic science to turn these findings into a path toward truly personalized medical nutrition.

My Healthy Diet Recommendations

This section offers my best dietary advice based on the research I have been sifting for decades and utilizing in my clinical practice with innumerable patients. In those few cases in which it is available, I also share the growing research regarding the epigenetic impact in each category below and how these mechanisms also relate to the function of the seven biochemical pathways. My chief recommendations are as follows:

Eliminate junk food. First of all, we need to eliminate those fast foods that distort and damage ordinary metabolism because of both the poor quality of their components and their toxic additives. This includes factory-processed or GMO foods of all kinds; sugar in its various forms; and empty calories from refined grains. The issue of junk food is well understood by my readers so I won't belabor it.

Detoxify your body. In *The Detox Prescription* (Rodale, 2015), Dr. Woodson Merrell underscores the fact that the human body has an extraordinary natural ability to detoxify itself. We rely on this efficient system when we wait for a hangover to lift or recover from a bout of food poisoning. But as I have noted, this native capability was not designed to handle the hundreds of toxic chemicals that now pollute our environment and our foods, especially if we are not eating organic products. Cutting-edge science is revealing how these pollutants can actually damage our genes and lead to conditions such as obesity, cognitive dysfunction, pain, allergies, infertility, and heart disease—all of which are on the rise. The good news is that we can supplement the work of this natural cleansing system with additional practices such as the detox method I recommend in Chapter 3. In his comprehensive book on this topic, Merrell shows readers how to assess their toxic risk factors and offers 100 nutrient-rich detoxification recipes that can be utilized in 3-, 7-, and 21-day cleanses. Of course, there are numerous other helpful detox methods that may be recommended by your own health providers.

Avoid refined sugar. As our genome evolved, we ate the equivalent of 22 teaspoons of sugar per year, or about one-fourth of a pound. These days the average American eats 150–180 pounds of sugar per year—or over half a pound of sugar a *day*! Clearly we are no longer eating in harmony with our genes.

In the last thirty years, our sugar calories have increasingly come from high-fructose corn syrup, mostly in the form of liquid calories from sodas, soft drinks, and sweetened beverages.

Please don't drink your calories in this reckless way! We need to "unjunk" our diets by eliminating high-fructose corn syrup as well as other toxins such as trans fats, artificial sweeteners, artificial food colorings, and MSG.

Processed sugar is unfit for human consumption. In fact, *the number one factor that leads to obesity and diabetes is sugar in all its forms.* Dr. Walter Willett from the Harvard School of Public Health stated at a recent White House meeting on prevention and wellness, at which I also spoke, that the two most important factors driving our obesity epidemic are sugar-sweetened drinks and the number of hours of television watched per day. These two behaviors, he pointed out, correlate more closely with obesity than any other factors in the research. The upshot is that we need to stop consuming so much sugar, especially when combined with a sedentary lifestyle. It's killing far too many of us.

Consume plenty of antioxidant nutrients. We're beginning to learn how antioxidants talk directly to our genes. As noted before, it's essential to remember that all fruits and vegetables with natural pigments are antioxidant, including red grapes, red cabbage, lemons, carrots, berries, oranges, green tea, and green vegetables. Among the berries, the most delicious, effective, and functional of foods in this category for reducing oxidative stress is blueberries. Among the specific nutrients that can reduce oxidative stress are vitamins A, C, and E, CoQ10, alpha lipoic acid, folic acid, zinc, manganese, beta carotene, and an overall reduction in saturated fat.

Stay clear of inflammatory foods. Foods that cause inflammation are coming into high focus these days, in large part due to the deleterious effects of factory farming on the quality of our food supply. This includes all factory-farmed meats, dairy, and grains—especially wheat products. That's why I ask all of my patients to eliminate gluten products as well as dairy for 2-3 weeks and see how they feel. A great many of them are allergic to gluten without knowing it; further, a small portion of the population is dairy-allergic without being aware, often because of genetic inheritance. In such cases, these foods can cause serious inflammation.

By now almost all of us have heard about the ravages of gluten, a substance found in bread and other stock foods made from wheat produced from industrialized agriculture. In *The Better Brain*, eminent neurologist Dr. David Perlmutter declares war on this all-too-common foodstuff, attributing a bewilderingly wide assortment of maladies to its consumption. He explains in detail how the industrialization of the farming of wheat crops has transformed a once-safe food into a terrible scourge. According to Perlmutter, dementia and many other brain diseases are not inevitable, nor are they genetic. They are directly and powerfully linked to a diet high in sugar and grains. Such a diet profoundly influences nerve health and brain function. His earlier book, *Grain Brain,* explains how the American diet rich in gluten and inflammatory foods is linked to neurological conditions. The book features health advice, a number of gluten-free recipes, and relevant case studies. Perlmutter's work explains why our brains are under siege with sky-rocketing rates of depression, dementia, ADHD, autism, and more that result from these processed foods.

Of course, many other features of our typical diets also contribute to inflammation. We've noted other culprits that include industrialized corn and soy products as well as our high consumption of red meats and other bad fats (including trans fats and partially hydrogenated fats). In addition, diets with a high glycemic load (such as potatoes, white rice, white bread, and sugary desserts) are also associated with inflammation because such carbohydrate-rich foods quickly raise blood glucose levels. If you combine the popularity of such unhealthy foods with our low consumption of the healthiest foods (fruits, vegetables, nuts, whole grains, insoluble fiber, and omega-3 fatty acids, etc.), you can readily understand the true cause of today's drastic rise in chronic inflammatory illnesses.

Still other dietary factors are associated with reduced systemic inflammation. Especially when we consume more variety (but not necessarily more quantity) of fruits and vegetables, more benefits in reduced risk for chronic inflammation diseases are seen in the research. In general, human clinical studies have found that higher fruit and vegetable intake is always associated with lower levels of inflammatory biomarkers. Newer studies are also showing a strong connection between our gut biome and inflammation.

Among the anti-inflammatory fruits and vegetables to always include in your daily diet are nuts and seeds, berries, extra virgin olive oil, garlic, and ginger; among the nutrients are turmeric, quercitin, resveratrol, omega-3 fish oil, flaxseed oil, and monosaturated fats.

Both quercitin and resveratrol in particular have a potent impact on turning down the genes that create excessive inflammation. Resveratrol is abundant in red wine, while quercitin is a *flavanoid* that is found in apples, onions, cherries, some citrus fruits, dark leafy greens, broccoli, raspberries, black tea, green tea, red wine, and red grapes. (Flavanoids are responsible for the vivid pigments found in many fruits and vegetables, as noted above.) Supplements to use include omega-3 products and antioxidant vitamins listed earlier. And be sure to limit exposures to pollutants in your food by eating organically and by drinking purified water.

What else can we do to eliminate inflammation? *Polyphenols* (versatile chemicals found in many fruits and vegetables including cocoa/chocolate and even coffee) are important, and dietary fibers (see next section) are also crucial biochemical agents that diminish inflammation. Other foods that are known to be anti-inflammatory include tomatoes, whole grains, mushrooms, beets, and avocado.

A great deal of research remains to be done to identify the specific components of the therapeutic activities of anti-inflammatory foods. But enough is known right now to change our daily dietary practices, beginning with the basic advice I provide above.

Eat a high-fiber diet. As our sugar consumption has increased, our fiber consumption has decreased. Fiber is important because it slows the absorption of sugar into the bloodstream from our gut and also reduces excessive cholesterol.

Fiber mostly enters the diet from fruits, vegetables, nuts, seeds, and beans. On average we now eat less than eight grams per day, but our Paleolithic ancestors ate an estimated one hundred grams of fiber per day. This is yet another indication of how we are no longer eating in harmony with our genes.

Those who eat a refined, processed diet that comes from boxes, packages, or cans get much less fiber than those who eat whole, real foods. This lack of fiber in the average diet has enormous implications for health, becoming a contributing cause of heart disease, diabetes, obesity, cancers, and many other chronic diseases. In fact, studies show that if a diabetic person adds high levels of fiber to their diet, this one dietary change alone is as effective for them as their diabetes medication, but without any of the side effects. We return to the issue of fiber below when we discuss metabolic syndrome.

Consider the benefits of moderate alchohol. Another positive recommendation, but one that is subject to more research, is to consume small amounts of alcohol. In particular, the health benefits of red wine have been well documented in recent years. Studies have revealed that those who drink a glass of red wine a day are less likely to develop dementia or cancer. And there is also evidence that it could help regulate blood sugar—which could mean that a daily glass of red wine could help people with type 2 diabetes keep their blood glucose levels under control. This is the case because red wine contains polyphenols, which in some studies is shown to help the body control its glucose levels. (While all types of wine contains polyphenols, red wine contains the highest levels.) Researchers have found that the polyphenols in wine bind to a molecule or receptor that's involved in the regulation of blood sugar. Those who claim that drinking wine is good for your health strongly suggest only a moderate intake of just one or two small glasses a day, which is best taken with a meal.

Other research shows that moderate drinkers in general live longer than nondrinkers and heavy drinkers. However, nothing is simple in the world of clinical nutrition. A study of 53,000 adults who participated in a United Kingdom health survey suggests that consumption was associated significantly with mortality benefits only in younger men and older women. "These results call into question the widely held assumption that moderate alcohol consumption confers health benefits in most adults," according the authors of this report.[25]

Other recommendations. Eat moderate protein with each meal, such as eggs, chicken, fish, nuts, seeds, beans, tofu, or soy. It should be noted, however, that the *phytoestrogen* in soy may pose a problem because of its biochemical similarity with human estrogen; many limit soy intake for that reason,

25 "Association between Moderate Alcohol Consumption and Mortality Benefits: Dwindling Evidence?" See http://www.jwatch.org/na37014/2015/02/26/association-between-moderate-alcohol-consumption-and.

but more research is needed. Once again, be sure to get good fats into your diet by including omega-3 supplements and/or foods high in omega-3 oils such as olive oil, sardines, wild-caught salmon, and extra virgin coconut oil. Exercise and be physically active every day—that should go without saying. Enhance your metabolism with energy-boosting supplements such as acetyl-L-Carnitine, alpha lipoic acid, coenzyme Q10, N-acetyl-cysteine, NADH, D-ribose, resveratrol, and magnesium aspartate.

The Challenge of Metabolic Syndrome

Metabolic syndrome, a condition that was once rare, now affects over one billion people worldwide. This disorder is most often associated with heart disease and type 2 diabetes, but people afflicted with it also face an increased risk of stroke, dementia, cancer, blindness, and kidney failure. According to a standard definition, metabolic syndrome represents a clustering of at least three of these five medical conditions: abdominal obesity, high blood pressure, high blood sugar, low levels of HDL (the "good cholesterol"), and high triglycerides (i.e., an excessive amount of a normal blood fat). Measures of each of these five elements are crucial biomarkers in several biochemical pathways, especially lipid metabolism.

Tragically, roughly a third of U.S. adults have the syndrome, and nearly half of those aged 60 and older suffer from it. (These statistics are based on data that was gathered from 2003 through

2012 for the *National Health and Nutrition Examination Survey.*) These high percentages constitute a serious threat to our nation's health that is likely to increase in the future, creating an unsustainable burden on our healthcare system.

Our current approach to prevention and treatment of metabolic syndrome is obviously not working, and we're now in a deep crisis. And so, you would think that the questions on everyone's mind would be: Why is this happening? What has caused this epidemic? Why have our diets and treatments failed us so miserably? What new approaches can we take to treat this problem effectively? But even if some of us are asking the right questions, we're not being given the right answers.

The reason, of course, is that our "disease-care" model of medicine avoids asking such questions because of commercial imperatives that too often are married to an obsolete paradigm of human biology. Sadly, our current approach to treating metabolic syndrome also fails the test of common sense because it focuses on treating the symptoms of the disease rather than the causes. We are heading into disaster because of an embarrassing logical fallacy—the tendency of allopathic medicine to elevate the effect (the symptom) over its cause.

Conventional doctors diagnose metabolic syndrome patients with many different disease names. These include insulin resistance, pre-diabetes, obesity, syndrome X, adult onset diabetes, and type 2 diabetes. However, these are all basically the same malady with varying degrees of severity. The underlying cause and its proper treatment are the same as well.

To date, the attention of disease-care medicine has long

been on providing treatments that lower blood sugar (insulin therapy), lower high-blood pressure (antihypertensive drugs), lower cholesterol (statins), and thin the blood (aspirin). Unfortunately, such doctors rarely ask the most important questions: *Why* is your blood sugar too high? *Why* is your blood too sticky and likely to clot? Again, the true scientist must always ask: What are the root causes of any phenomenon under observation?

Regarding metabolic syndrome, here is my own answer based on fifty years of clinical practice and research: Elevated blood sugar, high blood pressure, and bad cholesterol are downstream symptoms that result from the interaction of factors such as diet, lifestyle, stress, and environmental toxins with an individual's unique genetic vulnerabilities. Thus, the obvious solution for metabolic syndrome is a comprehensive diet, stress management, and lifestyle program like the one I have presented in this book. It especially requires a diet based on an understanding of the previous discussion of the seven pathways—and if available, feedback measurements of biomarkers of the sort recommended in my final chapter.

Putting a Stop to Diabetes Mismanagement

To illustrate how drastically our current healthcare system mismanages metabolic diseases, let's focus on the case of diabetes. As you know, the standard treatment for this disorder involves the lowering of blood sugar levels through insulin medication. How well does this extremely common protocol work, you may wonder? I am sorry to have to report that insulin therapy for diabetes not only doesn't work well—it actually *increases* your risk of death.

First published in the *New England Journal of Medicine* in 2008, the extraordinary findings of the ACCORD study revolutionized our understanding of the treatment of this all-too-common disease. This study is a profound piece of medical literature that provides startling evidence for why conventional treatments for diabetes simply do not work.[26]

The ACCORD researchers followed over 10,000 diabetics who received conventional drug therapy to lower blood sugar. These patients were monitored closely, and their incidence of heart attack, stroke, and death was measured frequently. Much to the surprise of the researchers, *the patients who had their blood sugar lowered the most had the highest risk of death*. How could this happen if, as we believe, elevated blood sugar is the *cause* of all the evils of diabetes? Why would lowering blood sugar lead to *worse* outcomes?

Amazingly, the study had to be stopped after three and a half years because it became obvious that aggressively lowering blood sugar led to more deaths and more heart attacks. This finding completely explodes the way conventional medicine understands and treats diabetes. It's a revolutionary study, and one that was long overdue. Yet for those of us who

26 See Mark Hyman, MD, "The Diabesity Epidemic Part II: Why Conventional Medicine Makes Things Worse," https://www.huffingtonpost.com/dr-mark-hyman/the-diabesity-epidemic-pa_b_389423.html (Mar. 18, 2010).

have been working to understand the *causes* of diabetes, it isn't all that surprising.

How could lowering blood sugar increase your risk of death? The reason is simple: Elevated blood sugar is not a cause of disease, but merely one symptom of underlying metabolic, physiologic, and biochemical processes that are out of balance. In other words, lowering blood sugar with medications does not address the underlying issues that gave rise to the high blood sugar in the first place. What's more—and this additional revelation may surprise or even shock you—many of the methods used to lower blood sugar such as insulin injections or oral hypoglycemic drugs may make the problem worse by *increasing* insulin levels.

Clearly, a better approach to diabetes is one that requires knowledge of nutrigenomics as well as the most advanced medical nutrition research. To begin with, addressing nutritional deficiencies is a particularly important way to prevent and treat diabetes. Typically, these deficiencies include vitamin D, chromium, magnesium, zinc, biotin, omega-3 fats, and antioxidants such alpha lipoic acid. These nutrients are necessary for proper control and balance of insulin and blood sugar, and they happen to be among the very substances that most Americans are deficient in. When they are very low in our diet, our biochemical machinery slows down and can even grind to a halt.

But what's most critical to understand is that the solution to the diabetes epidemic can be found at the end of your fork! That is true for one simple reason: Food is more than calories;

it is information. We generally think of food as a means to feed our bodies the fuel they need to function. However, the new science of nutrigenomics has shown, as we have noted, that food literally speaks to our genes, turning them on and off or up and down. This daily input provides your DNA with instructions about how to control your metabolism from moment to moment and day to day.

Ultimately, by feeding your body the right information, you can turn off the genes that lead to metabolic syndrome and turn on the genes that lead to health!

So what should you eat to prevent metabolic syndrome and such metabolic diseases as diabetes and obesity? It turns out that, in my considered view, the optimal diet to prevent and treat diabetes is the Mediterranean diet. As we have noted, this is a diet of whole, real, fresh food—one that requires that you obtain, prepare, and cook from the raw materials of nature. Along with this approach, my supplemental dietary recommendations listed above should be implemented.

Dr. Mark Hyman's pioneering books offer us additional treatment ideas in this connection. Among his specific recommendations are to take a sugar-busting supplement called *PGX*, which contains *glucomannan*, a dietary fiber made from Konjac root (an Asian yam). Hyman calls it a "super fiber." Glucomannan can also be obtained from special noodles called Shirataki.

Of course, many of these same lessons apply to other diseases as well. For example, over the last 20 years, Dr. Dean Ornish, a pioneer in medical nutrigenomics, has proved that you could reverse blockages in clogged arteries and increase blood flow in the heart by changing the quality of your diet and engaging in some simple lifestyle changes. He also showed that you could beneficially affect over 500 genes—that is, literally turn off the disease-causing genes and turn on the health-promoting genes—by changing dietary intake and lifestyle practices over a period of just three months. This result is more powerful than virtually *any* medication currently available.

We've also seen that other factors contribute to reduction of heart disease risk. After 25 years of follow-up in one study, cardiovascular-related mortality was 15 percent lower among those who ate the most whole grains, including even popcorn. Investigators examined data from two large prospective studies of a cohort of U.S. healthcare professionals that included about 118,000 adults who were free of cancer and cardiovascular disease at enrollment. Intakes of whole grains (e.g., wheat, oats, cornmeal, rye, bulgur, buckwheat, brown rice, popcorn) and other foods were assessed every 2 to 4 years for 25 years. If the association demonstrated in this study is indeed causal, then these results would point to a reduction in cardiovascular-related mortality. It is interesting to note that cancer-related mortality was not improved for the subjects in these studies, but this fact actually reinforces the validity of the cardiovascular results. If you are a clinician, you now have another reason to counsel patients to eat more whole grains. But in case you are encouraging greater intake of popcorn as part of this regimen, tell them to forget the butter!

Additional Support for Fish Consumption

Previously I noted the health benefits of adding more seafood to our diet, a choice that is supported to a surprising extent by U.S. government recommendations.[27] In 2010, the government advocated more fish consumption; but in 2015 it went a major step further in their support for the benefits of fish.

These most recent guidelines were formulated by a panel of fourteen experts. Overall, the panel concluded that eating diets higher in whole plant foods and lower in calories and meats are healthier for both people and the environment. Their general recommendations came in the form of three types of healthy dietary "patterns," as they call it:

- The Healthy U.S.-Style Pattern
- The Healthy Vegetarian Pattern
- The Healthy Mediterranean Style Pattern

Referring to these three classifications, the panel concluded, "A moderate amount of seafood is an important component of two of the three of these dietary patterns (i.e., the U.S. and Mediterranean patterns) and has demonstrated

---

27 Such dietary guidelines are devised jointly by the U.S. Department of Health and Human Services and the Department of Agriculture. The guidelines are used by health care professionals, policy makers, educators, and food marketers, and they guide federal education and food assistance programs. See "Scientific Report of the 2015 Dietary Guidelines Advisory Committee," https://health.gov/dietaryguidelines/2015-scientific-report/10-chapter-5/default.asp (Feb. 2015).

health benefits." They also observed that each of these two diets replaced refined grains with whole grains and reduced the intake of salt, saturated fat, and added sugars. The latter are replaced with whole plant foods, seafood, and low-fat dairy.

Also, this panel addressed the crucial issue of the alleged risks of heavy metal contamination in seafoods. To the surprise of some, the panel concluded that "for the majority of wild caught and farmed species, neither the risks of mercury nor organic pollutants outweigh the health benefits of seafood consumption." This was even the case with pregnant women—another surprise—and this determination remains true, said the panel, even if the women doubled their consumption of a high-mercury fish like tuna from 6 to 12 ounces a week. Clearly, the panel underscores the fact that consuming fish with its omega-3 fatty acids is very beneficial indeed. In addition, they advise that we consume a wide variety of types of seafood.

One of the panel's other controversial observations was that "farm-raised seafood has as much or more EPA and DHA [both are omega-3s] per serving as wild caught." But this statement may be misleading. Not pointed out in the report is that farmed salmon also has higher levels of total fat, including "bad fats" such as saturated fat, and omega-6 fat. Omega-6 is usually not desirable, since it blocks some of the benefits of omega-3 fatty acids and promotes inflammation. We must always bear in mind the distinction between "good" and "bad" fats in our diets.

In Search of the Optimal Diet

A growing number of randomized controlled clinical trials suggest that protein and fat (good fat) don't deserve to be de-emphasized, as they have been in some well-known diets—especially in the famed Dean Ornish diet, which recommends a very low-fat, high-carb, plant-based diet with limited protein.

A case in point is a 2007 clinical trial led by Dr. Christopher Gardner at the Stanford University School of Medicine. Gardner's research team randomly assigned over 300 individuals to four groups: One group was assigned the high-fat, high-protein, and low-carbohydrate Atkins diet; the second was assigned Ornish's vegetarian diet, which requires consuming fewer than 10 percent of calories from fat; the third was assigned the Zone diet, which aims for a 40/30/30 percent distribution of carbohydrate, protein, and fat; and the fourth was assigned the high-carbohydrate, low-saturated fat LEARN diet (which stands for: lifestyle, exercise, attitudes, relationships, nutrition). The participants all lost about the same statistically significant amounts of weight. Notably, the Atkins dieters saw greater improvements in blood pressure and HDL cholesterol than did the Ornish cohort.

The recent multicenter PREDIMED research study also supports the notion that "good" fat intake is essential. It found that participants who spent five years on the Mediterranean diet were about 30 percent less likely to experience serious heart-related problems compared with individuals who were told to avoid fat. Bear in mind that the Mediterranean diet provides about 40 percent of its calories from good fats.

Protein too, is important, especially when one considers the 2010 trial published in the *New England Journal of Medicine* that found individuals who had recently lost weight were more likely to keep it off if they ate more protein. Corroborating this finding is the 2005 *OmniHeart Trial.* This study reported that individuals who substituted either protein or mono-unsaturated fat for some of their carbohydrates reduced their cardiovascular risk factors compared with individuals who did not.

But as we have noted, not all animal proteins are equal. A 2010 systematic review and meta-analysis of 20 studies found consumption of processed meat—such as hot dogs or pepperoni—was associated with an increased risk of diabetes and heart disease, but eating unprocessed red meat was not. In addition, a 2014 meta-analysis similarly reported much higher mortality risks associated with processed meat compared with red meat consumption and found no problems associated with white meat. And it is worth noting that in a study of people over 65, heavy consumption of animal protein provided better protection against cancer and mortality.

A recent article in *Scientific American* argued against the claims Ornish makes about the success of his diet. His 1990 Lifestyle Heart trial involved a total of 48 patients with heart disease: 28 were assigned to his low-fat, plant-based diet, and 20 were given the usual cardiac care. After one year, those following his diet were more likely to see improvement in their atherosclerosis. In its critique, the article notes that the patients who followed his diet also were told to quit smoking, start exercising, and attend stress management training, while the control group could pursue its usual lifestyle practices. "It's hardly surprising that quitting smoking, exercising, reducing stress and dieting—when done together—improves heart health," writes the critic. "But the fact that the participants were making all of these lifestyle changes means that we cannot make any inferences about the effect of the diet alone. . . . It's possible that quitting smoking, exercising, and stress management, without the dieting, would have had the same effect—but we don't know; it's also possible that his diet alone would not reverse heart disease symptoms. Again, we don't know because his studies have not been designed in a way that can tell us anything about the effect of his diet alone. There's also another issue to consider: Although Ornish emphasizes that his diet is low in fat and animal protein, it also eliminates refined carbohydrates. If his diet works—and again, we don't know for sure that it does—is that because it reduces protein or fat or refined carbohydrates?"[28]

Ornish's low-fat, whole food, plant-based approach has many strong benefits when compared to the highly processed, refined-carbohydrate-rich diet of most Americans today. But Ornish was a leader in the movement to cut down on fat and

28 Melinda Wenner Moyer, "Why Almost Everything Dean Ornish Says about Nutrition Is Wrong," https://www.scientificamerican.com/article/why-almost-everything-dean-ornish-says-about-nutrition-is-wrong/ (June 1, 2015).

protein consumption in the 1980s and 1990s. As a result, tens of million of Americans began to eschew protein and fat, including good fats. To fill the gap, they drifted toward the embrace of high-sugar and high-calorie processed foods, not in small part because of the onslaught of food industry advertising in those years. Many researchers believe that this inadvertent outcome may be a critical reason for the obesity epidemic we see today.

The Human Microbiome and Nutrigenomics

While hopefully clarifying much for you, this chapter has also revealed the many remaining uncertainties in the current science of nutrition—even given the new advances in nutrigenomics. So you may wonder: What's right for me out of this wide array of dietary recommendations and sometimes contradictory studies? How can I create a truly personalized approach to my diet?

A very promising new area of individualized testing has emerged in the last few years that may reveal which specific practices are right for you. This approach comes from today's fascinating and sometimes paradigm-busting research on the human microbiome.

The hundreds of species of *microbiota* that live in our bodies contribute directly to both health and disease. They are located primarily along the lining of our intestines (also on the skin and in the mouth and vagina), but they enjoy an intimate biochemical relationship with *all* the cells and organs of our body. In fact, the microbiome is so prominent in our biology that its presence influenced the long-term evolution of human DNA and even of our organs, including the brain. That's because these microbes evolved symbiotically in the guts of hominids for hundreds of thousands of years, and before that lived in symbiosis with all living things for several billion years. These huge colonies of bacteria are an integral part of us, and some researchers even suggest that we are a part of *them*. Other researchers say that we are a kind of "human-microbe hybrid," a veritable superorganism.

As noted earlier, our microbiomes consist of more than 100 trillion bacteria that live in our intestines and outnumber our cells by 10 to 1. And, each of our microbiomes is genetically unique: Humans share about 99.5 percent of the same DNA, but the genetic profiles of our microbiomes vary widely. This should not be surprising when you consider that, while your body contains about 23,000 unique genes, your gut microbe consists of *over one million bacterial genes*. That has led some to suggest that it amounts to "a second human genome."

Among other tasks, these trillions of microbes in our gut determine how we digest and utilize everything we eat. What's more, there is a growing body of evidence that the condition of our microbiome is correlated with a wide range of diseases that include asthma, obesity, heart disease, cancer, Crohn's disease, autism, anxiety, Alzheimer's, and depression.

Our biome has much, very much to do with optimizing good health, and it is an especially prominent player in our immunity and inflammation pathways. For example, an April 2013 article in *Mother Jones* on our industrialized food system quotes one researcher who found that "literally within minutes" after eating a McDonald's breakfast, she was able to measure a large spike in the presence of *C-reactive protein* in the blood of her test subjects. This biomarker, you'll recall, is a solid indicator of systemic inflammation.[29] Furthermore, research is beginning to suggest that the most important variable in obesity isn't willpower or genetic susceptibility—and not even one's caloric intake. Instead, the chief cause may turn out to be the very composition of our gut bacteria, which determines which nutrients get extracted from food and in what quantity. The incidence of obesity (and many other diseases linked to the biome) may come down to whether a person consumes sufficient fiber, a crucial precursor to healthy gut flora. *The microbes in our gut feed on soluble fiber*, and its intake causes them to release many healthy substances such as B vitamins and vitamin K. And, we know that fast-food breakfasts contain almost *no* fiber.

The high-carbohydrate foods and bad fats that typify the American diet may indeed be killers, doing so via the gut biome. This diet promotes microbial imbalances that lead to the secretion by the gut of *endotoxins* and other inflammatory substances that "leak" into the bloodstream. These harmful molecules cause the familiar biochemical cascade in

---

29 This account is quoted in Chopra and Tanzi's *Super Genes*. The authors provide a useful overview of the microbiome, which they call "the new power player in health." See pp. 75–96.

the inflammation pathway that leads to obesity, type 1 diabetes, and many other chronic maladies linked to metabolic syndrome. This *gut-inflammation* link now seems to be crucial in our health, but the simple shift to more soluble fiber in the diet may be able to stop it in its tracks, according to many cutting-edge researchers. In fact, strong evidence shows that certain microbes directly generate epigenetic changes (known in this case as *histone modifications*) in the colon that protect it from cancer, provided that there is sufficient fiber in the diet.[30]

As I write this, however, the science of assessing the human microbiome is far from being as precise as that of genetic assays, which are increasingly precise. As a result, the commercial companies in this new field offer disclaimers stating that their profiles of your microbiome are for purely educational purposes and cannot be used yet to diagnose or predict your future health. But because of the exponential rate of change in this crucial field—not least of which will be advances in bioinformatics—these profiles will soon provide us with medically reliable biomarkers. In the meantime, many health practitioners have already begun to rely on assays of microbiomic data in their estimates of health prospects.[31]

---

30 Mike McCrae, "Bacteria Living in Our Gut Are Hijacking and Controlling Our Genes," https://www.sciencealert.com/microflora-promotes-epigenetic-crotonylation-histones-gut-epithelial-cells (Jan. 10, 2018).

31 For example, one start-up company that offers profiles on your microbiome for "edification" is uBiome in San Francisco. Their sampling kit contains a swab that is used to collect a small sample from your mouth, nose, gut, or genitals for testing. Your results are then compared to reference groups of vegetarians, Paleo dieters, people on antibiotics, heavy drinkers, and other comparable profiles. Once you see how your profile compares

Beyond the influence of diet and nutrition on the human biome is a new area of research indicating that physical activity and fitness also have a major direct impact. Researchers from University College Cork published a 2018 study in *Gut* that is the latest in their series of findings that exercise is an important factor in the complex relationship between the "microbiota," or bacteria of the intestinal tract, with diet playing an important complementary role. Drawing on their findings, the research team noted that "the microbiota of athletes has become adapted to the metabolic demands of vigorous exercise, but it is unclear to us the degree to which exercise per se or the dietary changes that accompany vigorous exercise or both contribute to this functional adaptation of the microbiota."[32]

Dr. Fergus Shanahan and his colleagues compared the microbiome of 40 professional rugby players with 46 controls. Of the total of 19,300 pathways that were examined, 98 were found to be significantly altered. There were many major differences; for example, the athletes' microbiome had more pathways that could be used for potential health benefits. These pathways included better synthesis of carbohydrates and antibiotics, higher levels of probiotic bacterium, which is inversely related to obesity and metabolic disorders, as well as short-chain fatty acid metabolites that push toward

---

to the reference groups you belong to, you can make actionable decisions on steps to take in modifying your diet to achieve a more optimal state of health.

32 Quoted in Diana Swift, "Pro Rugby Players Prove Athlete's Guts Are Different," *Medpage Today* (Apr. 8, 2018).

a leaner body type and have a positive impact on immunity, colon cells, and brain function.

These findings echo a pilot study performed in the United States in 2017 that also found that professional cyclists had a more diverse microbiome than healthy nonprofessional cyclists. Based on these studies and others, Dr. Shanahan stated, "We make no recommendations from our research except to say that exercise is probably the single most important thing that anyone can do for their health. A state of physical fitness appropriate to one's age and health is what one should aim for, and this should be achieved slowly over time."[33] Future research will try to establish the levels and duration of moderate exercise required by nonprofessional athletes to achieve an optimal state in the microbiota of the intestinal tract.

From such studies it is clear that an understanding of the complexity of the human biome and the interactions between the intestinal tract, diet, exercise, stress, and other external environmental factors remains to be determined. Today, the interpretation of the biochemical analyses of the human biome is akin to walking into a supermarket and examining the bar codes on products. All the information is in the bar codes, but we do not understand the meaning without a bar code reader. That is precisely what is needed in the future of human biomic research and its applications to our daily health and longevity.

---

33 "Quoted in W. Barton et al., "The Microbiome of Professional Athletes Differs from That of More Sedentary Subjects in Composition and Particularly at the Functional Metabolic Level," *Gut* (2018).

The Crucial Relationship of Diet to Gut Bacteria

For some years, studies in mice had suggested a link between the gut bacteria and diseases of every kind. Researchers had trouble, however, pinning down similar connections in humans, in part because it's difficult to make test subjects change their diets long enough to alter the gut microbes sufficiently to see an effect on health—or so researchers once thought.

But Dr. Peter Turnbaugh provided a turning point in this important field in 2009. This Harvard microbiologist was able to demonstrate in mice that a change in diet affected their microbiome *in just a day.* So he and Lawrence David, now a computational biologist at Duke University, decided to test the hypothesis that a changed diet could have an immediate effect on the human microbiome as well. They recruited ten volunteers to eat only what the researchers provided for five days. Half ate only animal products—bacon and eggs for breakfast; spareribs and brisket for lunch; salami and a selection of cheeses for dinner, with pork rinds and string cheese as snacks. The other half consumed a high-fiber, plants-only diet with grains, beans, fruits, and vegetables. For the several days prior to and after the experiment, the volunteers recorded what they ate so the researchers could assess how food intake differed.

The scientists isolated DNA and other molecules, as well as bacteria, from stool samples before, during, and after the experiment. In this way, they could determine which bacterial species were present in the gut and what they were producing. The researchers also looked at gene activity in the microbes.

Within each diet group, differences between the microbiomes of the volunteers began to disappear. The types of bacteria in the guts didn't change very much, but the abundance of those different types did, particularly in the meat eaters. Reporting their findings in *Nature Online*, Dr. David Turnbaugh and his colleagues reported that in four days, bacteria known to tolerate high levels of bile acids increased significantly in the meat eaters. (The body secretes more bile to digest meat.) Increases were seen in the meat eaters in one bacteria, *Bilophila wadsworthia*, that is associated with inflammatory bowel disease in studies of mice.

Gene activity, which reflects how the bacteria were metabolizing the food, also changed quite a bit. In those eating meat, genes involved in breaking down proteins increased their activity, while in those eating plants, other genes that help digest carbohydrates surfaced. According to Turnbaugh, "What was really surprising is that the gene [activity] profiles conformed almost exactly to what [is seen] in herbivores and carnivores." This rapid shift even occurred in the long-term vegetarians who switched to meat for the study, he says. "I was really surprised how quickly it happened."[34]

From an evolutionary perspective, the fact that gut bacteria can help buffer the effects of a rapid change in diet, quickly revving up different metabolic capacities depending

34 Elizabeth Pennisi, "Extreme Diets Can Quickly Alter Gut Bacteria," http://www.sciencemag.org/news/2013/12/extreme-diets-can-quickly-alter-gut-bacteria (Dec. 11, 2013).

on the meal consumed, may have been quite helpful for early humans. And this flexibility has strong implications for human health today.

But it also adds additional complexity. Whereas we earlier saw how the epigenome results from an indeterminate "cloud" of influences, the microbiome is tough to track because it changes so rapidly. It is more malleable than the epigenome, but that fact may also mean that these friendly bacteria are there, in part, to help us adapt all the more quickly—more than any other biological factor—to our ever-changing experience of the world.

In this very hot area of science, researchers worldwide are now looking into how, by adjusting diet, one can shape the microbiome in a way that can promote health. We will soon be at a point where we can make sensible dietary recommendations aimed at improving the microbiota in the gut. A full understanding of the human biome and its relationship with diet, exercise, and other lifestyle factors remains to be determined.

The Seven Biochemical Pathways and Your Food Choices

As with other key factors in our health—such as exercise, stress management, exposure to pollutants, and our beliefs—what we eat can turn our genes on or off, or can activate profound changes in our gut that affect the entire body. These gene expressions and biomic alterations, we now know, determine which biochemical pathways are activated or suppressed. These pathways affect all of our organs, our brain, our nervous system, and our entire mind-body balance, for better or worse. The interactions between the pathways are subtle, ongoing, and ever changing, but nutrigenomics is teaching us how to navigate this complexity and move toward optimal health.

Our next step is to use what we know about epigenetics and the microbiome to remove most of the speculation about what diet and supplements are best for you as an individual. In the near future, nutrigenomics will provide a more objective, scientific basis for answering such questions. What I find inspiring as I write this is that we are only a few years away from being able to know in a matter of hours or days whether specific health practices are working for each person. The challenge won't be a matter of finding "the" right diet for everyone, but rather one of discovering, through well-designed tests, the right combination of choices that are healthy for each person according to their specific epigenome. But with improved genetic and microbiomic testing a few years away from being widely available, why not get the jump on it now? It is reasonable to expect that a strategy that combines moderation and a good sense of proportion along with emphasis on the key components of the Mediterranean diet—fruits and vegetables, fish, oils, nuts, and moderate red wine consumption—should help your brain remain at its cognitive best for years to come, and keep your body vital and healthy well into your senior years.

5

MIND MATTERS

Turn Off Genetic Vulnerabilities by Reducing Stress

Because of its mechanistic and reductionist bias, mainstream medicine has always been slow to acknowledge that our minds play a vital role in our physical health. This was especially the case in the sixties and seventies, when there was an ongoing debate about whether or not our thoughts and feelings could directly influence our biochemistry. In those days, the materialists had the upper hand, but the rising tide of evidence eventually forced them to give ground. Toward the end of that era, my book *Mind as Healer, Mind as Slayer*, published in 1977 by Random House, was among the first to herald this once-elusive mind-body connection. In that book I defined stress as a condition that affects both mind and body, and I showed how it contributes to four major types of chronic illness: heart disease, cancer, arthritis, and respiratory illnesses. Along with my other colleagues in the emerging field of integrative medicine, I have been able to demonstrate through my subsequent research and writing that we can reduce our chances of getting those and other degenerative diseases if we commit ourselves to managing our stress effectively.

In fact, I went even further in my research. I discovered that some people had the ability to *master* the connection of their mind and body. As I followed the thread of research, I met and studied adept meditators who—under strictly controlled conditions—were able to demonstrate that they could exert a remarkable degree of control over pain, bleeding, and infection once they had achieved a meditative state in the laboratory. These people were the pioneers, the exemplars, for those of us who aspire to turn our minds into lifelong allies for health and healing.

Jack Schwarz, a Dutch meditation teacher, was perhaps the most impressive of these masters. He had learned to control his own pain and bleeding when he was tortured as a prisoner in a Nazi concentration camp. When my colleagues and I first began to study him under exacting laboratory conditions at the University of California School of Medicine in San Francisco, Jack seemed ordinary at first. He exhibited perfectly normal baseline responses to pain and a normal bleeding time when he was not meditating. Next we asked him to meditate in the controlled conditions of the lab. When I subjected him to pain during his meditation session, we were amazed to observe that his brain waves showed none of the electrical changes normally associated with pain. We then went so far as to ask him to subject himself to pain with a self-inflicted wound by pushing a sharpened, unsterile knitting needle entirely through his left bicep. We discovered to our surprise that he was able to reduce the bleeding by accelerating the time it took for his blood to clot.

In a later documentary film for the Canadian Broadcasting Corporation, Schwarz again pushed a sharpened knitting needle completely through his bicep without any display of discomfort. As astounding as that was to witness, he insisted that the real significance of his demonstration was that all of us—not just exceptional individuals—are able to wield such a profound influence over our body. That insight has remained a major theme in my work ever since.

Very few of us will become so adept that we can impale ourselves with knitting needles, nor would we want to of course! But I believe that we all have the mental, emotional, and spiritual capacity to fully manage the levels of stress we experience every day. When I first published my study of Jack Schwarz, it was controversial. Not everyone in the media or in the medical community accepted the idea that we could actually control our own nervous systems. But today the important truth that *all of us are born with a natural ability to self-regulate stress* is a cornerstone of modern medicine, or at least integrative medicine.

More recently, an unexpected and perhaps an even more

exciting chapter is unfolding. With the knowledge explosion in human genetics, the same debate is back again, but in a new form: Can our minds directly influence our genes? What's different this time is that genetic testing and retesting gives us an unprecedented ability to precisely monitor the biological effects of our mental and emotional states. In other words, this time we can prove—almost right away—that we can modify gene expression by cultivating relaxed, healthy, and positive states of consciousness and self-awareness. We don't need to wait for long-term studies or engage in a lifetime of meditation or other consciousness practices in a vague hope that we might one day succeed. All we need is a blood test to see exactly how our genetic biomarkers have changed in response to our current behavior.

Let's consider just a few examples of such influence among the many studies that pioneering researchers worldwide have carried out in recent years, beginning with a scientific look at the lifelong epigenetic effect of childhood stress.

Epigenetic Changes from Early-Life Trauma

Early-life stress matters—we have proof that it has a measureable epigenetic effect. Research shows that undue strain or abuse experienced during a child's development affects that young person's epigenome far into adulthood, altering patterns of stress response and often leaving the child with lifelong physical vulnerabilities or emotional disabilities that require treatment. These early epigenetic influences literally burn trauma into the brain and body, or what I prefer to call *the body-mind.* In brief, here are three remarkable instances of how "mind as slayer" can bring about negative genetic alterations resulting from childhood trauma:

- A study of 448 Dutch soldiers showed that early-life trauma changed more than 45,000 genes in the hippocampus, causing these men to be more vulnerable to emotional stress and PTSD later in life. (The hippocampus is thought to be the brain's center of emotion and memory, and the heart of the autonomic nervous system.)

- A study of 204 undergraduate women whose DNA was collected before and after a campus shooting had startling results. The specific epigenetic modifications created by this early trauma predicted which of these women would later suffer from PTSD-related symptoms as a result of experiencing or being in proximity to the shooting.

- In another study of early trauma, 25 people with a history of abusive childhoods who died by suicide were compared to 16 controls who had died suddenly but who did not have abusive histories. Genetic analysis focused on 23,551 hippocampal genes. The group that had experienced abuse showed extreme expression (either greater or lower methylation) in 362 neuronal genes. This result means that the functional ability of their genes to assess and respond to danger had been radically changed by child abuse, and may in turn have resulted in their suicide. This unique study suggested that childhood trauma can alter the expression of genes that regulate neuronal function.[35]

The Impact of Early-Life Trauma and Stress on Telomeres

Can early-life trauma affect our telomeres? A 2012 study at Duke University suggests that it can, corroborating earlier studies that have found that children who are physically abused or bullied tend to have shorter telomeres. As discussed earlier, telomeres are the biochemical structures at the tips of chromosomes whose shrinkage has been linked to aging and disease; as cells divide, these structures grow shorter, limiting the number of times a cell can reproduce.

Previous research had already identified an association between stress and accelerated telomere loss. Plus, shortened telomeres were *sometimes* shown to correlate with other health problems including aging, smoking, obesity, mental illness, heart disease, and chronic fatigue. Telomere erosion has also been related to both oxidative stress and inflammation, but such links are not always present in telomere research. "There's a lot of doubt in the field," says Dr. Joao

35 B. Labonté et al., "Genome-Wide Epigenetic Regulation by Early-Life Trauma," *Archives of General Psychiatry* (July 2012) 69:722.

Passos, a cellular aging specialist at Newcastle University. "For as many studies that show telomere length as a good predictor of health outcomes, there are many that find no relationship."[36]

The Duke University study was more advanced in design than many previous efforts and led to a distinctive outcome. The Duke researchers used data from the Environmental Risk Longitudinal Twin Study, which followed British twins from birth. The team selected 236 of these children, half of whom had experienced at least one form of violence. Using DNA samples collected at ages 5 and 10, the investigators assessed how many times a particular gene had copied itself. Significantly, they found that gene replication was indeed lower among children who had experienced violence. The team not only noted a clear relationship between violence and shortened telomeres, but it also discovered a significant association between the *number* of violent experiences and the amount of telomere loss. According to one coauthor of the study, Dr. Avshalom Caspi, "Children who experience physical violence appear to be aging at a faster rate."[37]

Parents and caregivers may well wonder whether this process can be reversed once it is set in motion in childhood. Some studies suggest that making healthful lifestyle changes,

---

36 http://www.sciencemag.org/news/2012/04/childhood-stress-leaves-genetic-scars.

37 I. Shalev et al., "Exposure to Violence during Childhood Is Associated with Telomere Erosion from 5 to 10 Years of Age: A Longitudinal Study," *Molecule Psychiatry* (May 2013), 18:576, https://www.nature.com/articles/mp201232.

such as reducing stress, eating well, and exercising, can slow down the rate of telomere loss. But much more research is needed in this important field.

Altering the Expression of Prostate Cancer Genes

There is little question that early-life trauma has a direct and enduring impact on our genes, but it's not a one-way street. Just as mental stress and adversity can affect our genes, so also is it possible to have a *positive* impact through the mind. That is when "mind as healer" becomes a reality. In fact, some research shows that intervening to reduce stress through the development of better habits can push our genes toward healthy expression and directly enhance our wellness and longevity.

For example, in part because of this insight about the biological influence of the mind, we have recently witnessed a sea change in the diagnosis and treatment of prostate cancer. In 2012, the U.S. Preventive Services Task Force advised against the routine prostate cancer screening that provides doctors with PSA readings, since positive results on this test often lead to premature and excessive medical interventions with major negative side effects. Significantly, this advisory was based on the finding that the onset of prostate cancer is reversible through lifestyle changes.

Along this line, one new direction in the treatment of prostate cancer involves an intensive lifestyle and nutrition program focused on influencing the genes involved in prostate cancer. In 2008, a group of researchers at the UCSF School of Medicine in San Francisco enrolled 31 men with a low-risk form of prostate cancer who agreed to decline immediate surgery, hormonal therapy, or radiation while undergoing careful surveillance for the progression of their prostate tumors. Instead of the usual medical interventions, these men undertook an intensive, three-month program in which they followed a low-fat, plant-based diet and engaged in stress management practices. When comparing their PSA readings at the start and completion of the program, the researchers found these practices had decreased the expression of the genes associated with prostate cancer.[38]

Dramatic Genetic Changes from Massage

Direct body-mind interventions by caregivers can also positively change gene expression. For example, new research has revealed for the first time that the kneading of sore muscles by a massage practitioner can turn off genes associated with inflammation and turn on genes that help muscles heal. A unique study based on this hypothesis was designed and led by Dr. Mark Tarnopolsky, a neurometabolic researcher at McMaster University in Canada. Tarnopolsky had suffered a severe hamstring injury in an accident and had received

38 https://www.ucsf.edu/news/2015/11/253051/working-sweat-may-protect-men-lethal-prostate-cancer.

massage therapy as an essential part of his rehabilitation regimen. The massage therapy he received seemed to be so effective that Dr. Tarnopolsky set out to investigate the biochemistry behind it. He was surprised to find that, despite the widespread popularity of massage, researchers in that field knew surprisingly little about its molecular and genetic effects.

Two key benefits of massage had previously been well-documented: an increase in blood circulation in the massaged areas, and the general release of *endorphins* (which decreases pain and increases sensations of pleasure). Researchers knew these positive results were possible, but no one had yet explained how and why these effects occur. Dr. Tarnopolsky wanted to answer these and other questions.

Tarnopolsky and his colleagues designed and conducted their own study, recruiting 11 young men willing to undergo a grueling upright cycling session that left their muscles damaged and sore. Ten minutes later, a massage therapist massaged one of their legs. The researchers took tissue samples from the quadriceps of both legs of each of the volunteers at three points: once before the workout, once ten minutes after the massage, and once three hours after the workout. Then they compared the genetic profiles of each sample.

Samples taken before the massages but after the exercise were not a surprise. Researchers detected a greater presence of cell repair activity as well as acute inflammation in the post-workout samples than in the pre-workout samples; this was consistent with the established fact that exercise activates genes associated with these two processes. But what did surprise them were the clear differences between the massaged legs and the unmassaged ones after the exercise. Massaged legs showed 30 percent more expression of a gene that helps muscle cells build mitochondria, the cellular engines that turn a cell's fuel into energy. Even more impressive, these men also had *three times less* amount of a chemical that turns on genes associated with inflammation.

Dr. Tarnopolsky's results prove that massage reduces inflammation caused by exercise and promotes faster healing of affected areas. Incidentally, the study found no evidence to support widely believed claims that massage removes lactic acid, a by-product of exertion long blamed for muscle soreness. Most important, this study underscores the truly amazing finding that the human touch during massage actually induces positive changes in gene expression, leading directly to muscle healing and improved body-mind health.

Managing Gene Expression in Schizophrenia

In recent decades, geneticists have discovered that schizophrenia is highly heritable. But the small size of some studies and the immaturity of the field of bioinformatics left doubt about the specifics of the perceived link between genes and this mental disease.

This dilemma was remedied in 2015 by a huge genome-wide association study of nearly 37,000 cases of schizophrenia and 113,000 controls. The GWAS researchers identified 108 genetic loci linked to schizophrenia, 83 of which had not been reported previously.[39] (A loci, as noted earlier, refers to the fixed position—on a chromosome—of a gene or an epigenetic marker.) Most of these loci were located in brain cells that interact with dopamine or other neurotransmitters; intriguingly, some of these same gene locations are also involved in the body's immune response, which the scientists considered to be an important correlation because of previous research.

Many new potential therapeutic targets for nutrients or drugs are believed to have been identified because of these results. Further, the large size of this landmark study and the researchers' use of advanced statistical techniques lend weight to these findings. According to an article in *Nature* by the Schizophrenia Working Group of the Psychiatric Genomics Consortium, "This study supports the hypothesis that the biology of schizophrenia involves changes in neurotransmission that are affected by the acquired immune response. It also suggests that environmental agents might trigger the disease in individuals with genetic susceptibility since birth."[40]

---

39 Schizophrenia Working Group of the Psychiatric Genomics Consortium, "Biological Insights from 108 Schizophrenia-Associated Genetic Loci," *Nature* (July 24, 2014): 511:421, http://dx.doi.org/10.1038/nature13595.

40 Ibid.

The Epigenetics of Phobias

It has long been known that anxiety disorders can result from inherited genetic susceptibilities. Research that builds on this fact is now looking at the epigenetics of specific phobias.

For example, one research group showed that a single gene might actually account for claustrophobia. In a 2012 study of this phobia, the researchers focused on the GPM6A gene, which was already known to be responsive to stress. Previous research had linked this gene to regulation of opioid receptors and serotonin transporters and indirectly to the so-called human panic response as well as to depression in schizophrenia. The research team went further, finding that claustrophobic individuals had significantly more abnormalities in this gene when compared to nonclaustrophobic individuals.[41]

In addition to this discovery, pioneering research points to the possibility that some phobias might actually be the result of "memories" passed down by means of transgenerational epigenetic inheritance from ancestors who had suffered from the same phobia.[42] (We introduced epigenetic inheritance in Chapter 1.) This finding contrasts with the belief long held by psychologists that phobias can only result from traumatic personal experiences in a person's own childhood.

A remarkable test of the hypothesis of the epigenetic

---

41 A. El Kordi et al., "A Single Gene Defect Causing Claustrophobia," *Translational Psychiatry* (Apr. 30, 2013): 3:e254, http://dx.doi.org/10.1038/tp.2013.28.

42 R. Gray, "Phobias May Be Memories Passed Down in Genes from Ancestors," *Telegraph Sun* (May 24, 2015), 1–3, http://www.telegraph.co.uk/news/science/science-news/10486479/Phobias-may-be-memories-passed-down-in-genes-from-ancestors.html.

inheritance of a phobia was carried out by Dr. Brian Dias, a psychiatrist at the Emory University School of Medicine. Dias trained mice to fear the smell of cherry blossoms using electric shocks; later on they were permitted to breed.[43] To the great surprise of many, two subsequent generations of these mice showed fearful responses to this odor compared to a neutral odor, despite never having encountered these smells before. Autopsies of the brains of the trained mice and their offspring showed structural changes in regions that govern the sense of smell. Dias and his team concluded that the DNA of these animals carried epigenetic modifications on the associated genes.

"Our results," said Dias, "allow us to appreciate how the experiences of a parent before even conceiving offspring markedly influence both structure and function in the nervous system of subsequent generations. . . . Such a phenomenon may contribute to the etiology and potential intergenerational transmission of risk for neuropsychiatric disorders such as phobias, anxiety, and post-traumatic stress disorder."[44] Professor Marcus Pembrey, a pediatric geneticist at University College of London, strongly supports this controversial notion of the biological transmission of a phobia. "It is high time public health researchers took human transgenerational responses seriously. . . . I suspect we will not understand the rise in neuropsychiatric disorders or obesity,

43 https://www.nature.com/articles/nn.3594.

44 Ibid.; Gray, "Phobias May Be Memories Passed Down in Genes from Ancestors."

diabetes, and metabolic disruptions generally without taking a multigenerational approach."[45]

This and other research vividly points to the idea that transgenerational inheritance exists and is mediated by the epigenome. But much more research is needed to refine these insights.

Stress Hormones Can Cause Epigenetic Changes

For both mice and men, science is revealing the genetic and biochemical pathways through which chronic exposure to stress hormones can change gene expression.

During stressful situations, we produce beneficial hormones called *glucocorticoids* (GC) that affect many bodily systems. GCs are anti-inflammatory, but they also work in the body's immunity pathway. Because this class of hormones is known to be immunosuppressive, drugs containing pharmaceutical versions of this hormonal family are sometimes used to treat diseases caused by an overactive immune system, such as allergies and asthma.

Then again, too much of a good thing can cause new problems. Many past studies have found that an *excessive* amount of glucocorticoids can alter gene expression in the brain. And some researchers have suggested that such

45 Ibid.; Gray, "Phobias May Be Memories Passed Down in Genes from Ancestors."

influences extend even wider, in part because the distribution of the effects of GCs are mediated by what is known as the *hypothalamic-pituitary-adrenal axis* (HPA)—a network that links the hypothalamus and the pituitary gland in the brain with the adrenal glands near the kidneys. Thus, a more advanced question concerns how GCs affect the genes that regulate the entire HPA axis. A group led by Drs. James Potash and Gary Wand at the Johns Hopkins University set out to answer this question by testing the hypothesis that hormones affect the entire HPA axis through epigenetic modification.

The researchers added *corticosterone* (a glucocorticoid secreted by the adrenal gland) to the drinking water of mice for four weeks. After exposure, and again after a four-week recovery period without corticosterone, the scientists examined the expression levels of five HPA-axis genes. In particular, they measured the degree of methylation in each gene (a common form of epigenetic modification, as previously noted).

In the September 2010 issue of *Endocrinology*, the researchers reported that mice given corticosterone exhibited an altered expression of three of the five HPA-axis genes, which they attributed to decreased methylation.

Now, it turns out that methylation in these same genes has also in the past been associated with PTSD and mood disorders. These newer findings by the Johns Hopkins researchers suggest that epigenetic modification (through methylation) occurs because of the excessive secretion of the GC hormone. "This gets at the mechanism through which we think epigenetics is important," says Potash. "Epigenetic marks added to DNA through life experience may prepare an animal for future events. If you think of the stress system as preparing you for fight or flight, you might imagine that these epigenetic changes might prepare you to fight harder or flee faster the next time you encounter something stressful."[46]

Those of us who face stressors such as unreasonable work deadlines are—unlike our cousins in the animal world—unable to fight or flee, and such chronic stress may lead these emotionally trapped humans to experience a variety of hormonal disorders that are then "baked in the cake" epigenetically. This new research suggests that epigenetic changes may play a key role in creating stress-related diseases, and could point to effective treatments.

Meditation Positively Alters Gene Expression

Today's popular mindfulness and meditation practices come in many forms and are now known to have more than a few measurable effects on mental health. For example, the secular meditation-training program known as *mindfulness-based stress reduction* (MBSR) has long ago been shown to reduce depressive symptoms.

More recently, MBSR researchers at Duke University

46 https://www.nih.gov/news-events/nih-research-matters/stress-hormone-causes-epigenetic-changes.

wanted to know how the beneficial effects of this practice on depressives might vary according to demographics. For example, what happens if we factor in a person's religious belief system? Or, how does age or gender modify the effect of MBSR on depression?

To answer such questions, a team of researchers led by Dr. Jeffrey M. Greeson of the Duke Integrative Medicine Center studied the variations in depressive symptom outcomes among 322 adults who enrolled in an eight-week MBSR program. Remarkably, they discovered that depressive symptom severity decreased significantly, showing statistically significant reductions across all the identified subgroups—including religious affiliation, intention for spiritual growth, gender, and baseline symptom severity.

The discussion of their findings by Greeson and his colleagues is worth paraphrasing in detail:

The current results suggest that changes in depressive symptoms following MBSR are explained, in part, by increased mindfulness of thoughts and feelings and by an enhanced perception of spirituality in daily life. Given the connection between spirituality and mental health, mindfulness practice could parallel religious and spiritual practices, such as prayer and meditation. . . . Other MBSR outcome studies have reported that reduced depressive symptoms may be partially explained by lower levels of rumination, a known risk factor for depression. . . . Equally likely, the decrease in depressive symptoms may arise from the practice of disengaging from depressive thoughts and recognizing that they are just mental events rather than truth—a core skill called decentering.[47]

Because of this significant link to the reduction of depression, I would suggest that the next logical step for Greeson's team should involve research into the epigenetic effects of MBSR.

In fact, such an epigenetic link has already been discovered in a related form of meditation. The well-known form of meditation known as the *relaxation response technique* has recently been shown to not only have measurable psychological benefits but also discernible effects on gene expression.

Nearly 40 years ago, Dr. Herbert Benson of Harvard Medical School identified a discrete mind-body process that he named the *relaxation response* and showed it to be the physiologic opposite of the well-known *fight-or-flight response*. Benson describes it as a state of deep rest attained through breathing, meditation, yoga, and related practices. His so-called relaxation-response meditation technique is now widely used to help patients manage a variety of medical conditions from anxiety and chronic pain to cancer.

In an important additional breakthrough in 2013, researchers led by Benson at the Massachusetts General Hospital and Beth Israel Deaconess Medical Center reported that the relaxation response triggers changes in gene expression that can

---

47 Jeffrey M. Greeson, Moria J. Smoski, Edward C. Suarez, Jeffrey G. Brantley, Andrew G. Ekblad, Thomas R. Lynch, and Ruth Quillian Wolever, *Journal of Alternative and Complementary Medicine* (March 11, 2015) 21(3): 166–174, doi:10.1089/acm.2014.0285, https://www.ncbi.nlm.nih.gov/pmc/articles/PMC4365440/.

affect the body's immune function, energy metabolism, and insulin secretion.

One of Dr. Benson's collaborators at Beth Israel Deaconess is Dr. Towia Libermann, the co-senior author of the study. According to Libermann, the evidence arising from their study clearly links the relaxation response to *rapid* changes in gene expression. Libermann reported that genes involved both in immune disturbances and inflammation pathways were repressed after participants practiced the relaxation technique, while another set of pathways involved in mitochondrial function and energy production were epigenetically enhanced.[48]

In the study, 26 participants were longtime practitioners of the relaxation response and 26 others who had never experienced it before were trained in the technique. Researchers used gene profiling to identify changes in these subject's gene expression. "These changes," Libermann said, "occurred in both groups but were more pronounced among the long-time relaxers." Libermann, who had worked with Benson for the previous decade, says he was drawn to this research "to convince myself that there's really something going on here, and it's not just a placebo effect. . . .I'm [now] pretty convinced." He and Benson are currently investigating whether the relaxation response triggers molecular-level changes in people with hypertension, inflammatory bowel disease, irritable bowel syndrome, and other diseases.[49]

In a 2008 study that had focused on the long-term practice

48 http://journals.plos.org/plosone/article?id=10.1371/journal.pone.0062817.

49 http://commonhealth.legacy.wbur.org/2013/05/genes-altered-after-relaxation-practice.

of the technique, Benson and Libermann also discovered changes in stress-response genes. Blood samples from participants were analyzed to determine effects on 22,000 genes, and revealed significant changes in the expression of several important groups of genes over time. In particular, genes that manage energy metabolism were found to be upregulated because of the relaxation response, and pathways controlled by the activation of a protein called NF-κB—which is known to play a prominent role in inflammation, stress, trauma, and cancer—were suppressed by the practice.

Because of these positive results, Benson and Libermann have concluded that "relaxation causes multiple gene-expression changes that create 'mitochondrial resilience' by stabilizing key cellular processes during the adaptation to oxidative stress and by enhancing cell survival and function. The rapidity of these changes is noteworthy, as is the finding that more changes occur with more practice."[50]

Building on these results, researchers have demonstrated with convincing evidence that mindfulness meditation can induce *immediate and direct* modification of gene expression. Previous studies had shown dynamic epigenetic responses to diet or exercise within just a few hours, but for the first time, two new studies have demonstrated evidence of epigenetic changes following a single period of mindfulness practice.

The first is a landmark study published in 2014 in *Psychoneuroendocrinology* by researchers in Wisconsin, Spain, and

50 Ibid.

France that provides evidence of gene changes following a daylong period of mindfulness practice. After eight hours of disciplined sitting, meditators showed a range of genetic and molecular alterations, including reduced levels of pro-inflammatory gene expression that were not observed in the non-meditating control group.

"Most interestingly," writes Dr. Perla Kaliman of the Institute of Biomedical Research in Spain, who was the first author of the article, "the changes were observed in genes that are the current targets of anti-inflammatory and analgesic drugs. . . . Our findings set the foundation for future studies to further assess meditation strategies for the treatment of chronic inflammatory conditions." Another leader of the study, Professor Richard J. Davidson, founder of the Center for Investigating Healthy Minds at the University of Wisconsin-Madison, noted, "We can think of genes possessing a molecular volume control that ranges from low to high that will govern the extent to which the gene produces the protein for which it is designed. The genes that we found to be down-regulated with mindfulness mediation practice are those implicated in inflammation."[51]

Just as remarkable is a 2013 study from Harvard Medical School that showed immediate effects from a far briefer period of meditation.

In this study, gene profiles were analyzed in 26 long-term meditators before and after a mere 20-minute practice

---

51 https://www.psychologytoday.com/blog/the-athletes-way/201312/meditation-has-the-power-influence-your-genes.

session. These profiles were compared to 26 novices who were not meditators. In the long-term meditators, there was an increased expression of genes involving energy metabolism, mitochondrial function, insulin secretion, and repair of telomeres. Additionally, the genes involved in inflammatory response and oxidative stress were suppressed or turned off. Researchers were amazed that such changes could take place after only 20 minutes of practice by skilled meditators. Clearly, the changes in the state of consciousness in the minds of the meditators created a cascade of biological events that led to altered gene expression, which in turn changed specific biochemical pathways that govern their health and illness.

With these important studies, we now have compelling evidence of the power of the mind to move our genes and body chemistry toward optimal states of health and longevity.

The Pioneering Insights of Bruce Lipton

Dr. Bruce Lipton, formerly a research microbiologist and professor at the University of Wisconsin Medical School, goes yet another step further in his claims about the relationship of human consciousness and our epigenome.

This renowned and pioneering author of *The Biology of Belief* has long argued that our epigenome is altered by our ordinary perceptions and life experiences—which in turn are controlled by our beliefs. According to Lipton, genes can display a large degree of plasticity because our perceptions directly affect our blood chemistry, which in turn alters gene expression. But, says Lipton, another factor is even more important: Our perceptions and experiences *are shaped far more by unconscious beliefs than most of us think possible.*

Our mind, according to Lipton, always seeks unity and consistency. "The function of the mind is to create coherence between our beliefs and the reality we experience," writes Lipton. "What that means is that our mind will adjust the body's biology and behavior to fit with our beliefs." Again, the key factor in this equation is that the most powerful beliefs are unconscious. Ultimately, the state of our biology comes down to how the subconscious mind—which contains our deepest beliefs—has been programmed in childhood.[52] In this paraphrase of a recent interview, Dr. Lipton summarizes this point:

The major problem is that people are aware of their conscious beliefs and behaviors, but not of subconscious beliefs and behaviors. Most people don't even acknowledge that their subconscious mind is at play, when the fact is that the subconscious mind is a million times more powerful than the conscious mind and that we operate 95 to 99 percent of our lives from subconscious programs. . . . Your subconscious beliefs are working either for you or against you, but the truth is that you are not controlling

52 https://www.brucelipton.com/resource/article/epigenetics.

your life, because your subconscious mind supersedes all conscious control. So when you are trying to heal from a conscious level—citing affirmations and telling yourself you're healthy—there may be an invisible subconscious program that's sabotaging you.[53]

The upshot is that our subconscious mind determines our biology and chemistry, which changes the gene expression of our cells. "Cells are like miniature people [in the sense that they have] similar functions, including digestive, reproductive, immune, and nervous systems. Each cell, like every human, has receptors built into its skin so it can become aware of the environment. If a person is in a stressful environment, every one of our cells is also experiencing that stress via the electromagnetic vibrations sent throughout our body [by the stress response]. Similarly, if we are happy, our cells are happy and in harmony."[54]

Psychologists generally agree that our habitual patterns of behavior are programmed from childhood up until around the age of six, after which our life is more or less controlled by those habits. If in our childhood we suffered from negative influences, we will need to make a conscious effort in adulthood to unlearn how we were programmed to think and behave as a child and to not rely on such maladaptive habits. According to Lipton, there are three ways that are effective in changing, limiting, or sabotaging beliefs in the subconscious

53 Ibid.

54 Ibid.

mind: mindfulness meditation, clinical hypnotherapy, and a new healing modality known as "energy psychology."[55]

Two Effective Stress-Management Techniques

Based on the research we have sampled in this chapter, there is no doubt that meditation and stress management techniques offer benefits in developing a healthy epigenome. These methods are clearly a means to directly and positively influence our genes. Further, it appears that our cells swim in an "ocean" of biochemical influences that are governed, often negatively, by our perceptions and beliefs—which are in turn deeply conditioned by childhood programming. We can't see these issues because they lurk in our unconscious "shadow" self, and many of us remain in such denial of these effects that we fail to seek psychological or spiritual healing. Further, research shows that meditation practice alone may not solve these more intractable personal issues.

But for the purposes of this introductory chapter, let's turn back to the proven benefits of the simple practice of meditation or stress management, one of Lipton's three methods

---

55 According to Doc Childre, the founder of a popular energy psychology technique known as HeartMath, "'an energetic connection or coupling of information' occurs between the DNA in cells and higher dimensional structures—the higher self or spirit. 'The heart serves as a key access point through which information originating in the higher dimensional structures is coupled into the physical human system (including DNA), and states of heart coherence generated through experiencing heartfelt positive emotions increase this coupling.'" See Christina Sarich, "How Your DNA Is Affected by Quantum Intelligence," wakingtimes.com (Feb. 16, 2018), https://tinyurl.com/y9t84otg.

of psychological deprogramming. For those who desire to look into hypnotherapy and energy psychology, I refer you to Lipton's writings.

At a minimum, I often teach my patients two stress-management techniques they can practice anywhere and anytime. One focuses on breathing, while the other—known as *autogenic training*—is focused on the experience of heaviness and warmth. These techniques allow them to manage stress to attain a balanced, regenerative state that maximizes the efficiency of the seven pathways. Under ideal conditions, the pathways naturally work well, but as we have seen, daily stress, anxiety, or depression can interfere with their proper function. The truth is that meditation is actually deceptively simple, as you will see below. So, inspired by the profound findings of recent epigenetic research regarding such techniques, let's take a few minutes for some basic instructions. The form of mindfulness practice I recommend has been the basis for virtually all of the research I have cited. Those of you who have a solid meditation practice may skip this next section, but the following section on autogenic training is likely to be new to most readers.

Mindfulness Meditation—How to Get Started

Find a good spot in your home or apartment, ideally one with a minimum of clutter that affords some quiet. Leave the lights on or sit in natural light. You can even sit outside if you like, but be sure to choose a place with little distraction. Any quiet place is good, even a park bench or a sand dune at the beach.

At the outset, it helps to set the amount of time you are willing to commit by simply using a kitchen timer or the timer on your cell phone. Otherwise, you may occupy yourself over deciding when to stop. If you're just beginning, it can help to choose a very short time in your early sessions. Even starting with one minute is fine! Eventually you will be able to sit for ten minutes, and at some point twice as long as that. Ultimately you may decide to meditate for up to 40 minutes or even an hour.

I recommend that you do one session in the morning and one in the evening. But if you feel that your life is too busy for a morning and evening session of, say, twenty minutes—doing one meditation each day is better than none at all, even if it is short.

A proper meditation requires that you assume a good upright posture in a chair or on some kind of cushion on the floor. If you prefer the floor, you can start with a pillow or a folded blanket, and later on you might purchase a meditation cushion that will last you a lifetime. Sit with your feet flat on the floor if you are in a chair, or loosely cross-legged in lotus posture if you are on the floor. Any comfortable position is fine, but make sure you are stable and erect. If the constraints of your body prevent you from sitting erect, find a position you can stay in for a while. All that said, what follows is some basic instruction:

Straighten but don't stiffen your upper body. There is a natural curvature to the spine, so just relax until you feel that soft curve in your lower back. Your head and shoulders should feel balanced so that they rest comfortably at the top of your spine; it should not feel like you are holding them there, but rather balancing them. Keep adjusting your posture until it feels comfortable without holding yourself rigidly or with tight muscles.

Now, while maintaining your best posture, begin to pay attention to your breathing. Many meditation teachers say to follow your breath as you breathe in and breathe out, while others say to put more emphasis on the outbreath.

Inevitably, your attention will leave your breath and wander to thoughts, sensations, and distractions. When you get around to noticing that you have wondered off, simply return your attention to the breath. Don't bother judging yourself for forgetting your breathing or for getting concerned over the content of your thoughts. Just come back to following your breathing as your primary focus of attention. You'll often fall away from your focus, but you can gently come back. Soon you will settle into a relaxed state of mind in which you are simply witnessing the thoughts and feelings that arise while maintaining a single-pointed concentration on your breath.

This, then, is the most basic form of mindfulness practice. It is amazingly simple, but it's not necessarily easy to maintain. Our real work is to just keep doing it often, many times per week. Results will accrue, both in your sense of well-being and concentration and in terms of positive epigenetic changes.

The Technique of Autogenic Training

Autogenic training is a relaxation technique developed by German psychiatrist Dr. Johannes Heinrich Schultz in the 1930s. The technique can be used to alleviate many types of stress-induced psychophysiological disorders. This method involves the daily practice of sessions lasting several minutes that are usually done in the morning, but can be practiced at any time. During each session, the individual carries out a simple set of self-hypnosis instructions to induce a state of relaxation. They learn to bear in mind two key words—"heavy," which is the subjective sensation of muscles relaxing, and "warm," the sensation of increasing blood flow to the hands and feet, an indicator of overall relaxation.

Each session can be practiced in a position chosen from a set of recommended postures that include lying down or sitting upright, or any other relaxed and balanced position.

Autogenic training was popularized in North America particularly by Dr. Wolfgang Luthe, who coauthored, with Schultz, a multivolume tome on autogenic training. According to Wikipedia, in 1963 Luthe discovered the significance of "autogenic discharges," which he described as "paroxysmic phenomena" of a motor, sensorial, visual, and emotional nature related to the traumatic history of the patient.

There are many parallels between autogenic training and mindfulness meditation, both in its practice and in their positive outcomes. Here's a basic instruction for getting started:

• Find a quiet place free from distractions. Lie on the floor or sit relaxed in a chair. Loosen any tight clothing and remove glasses or contacts. Rest your hands in your lap or on the arms of the chair.

• Take a few slow, even breaths. If you have not already, spend a few minutes practicing breathing slowly and deeply into your diaphragm.

• Silently or quietly say to yourself, "I am completely calm."

• Focus attention on your arms. Silently or quietly and slowly repeat to yourself six times, "My arms are very heavy." Then silently or quietly say to yourself, "I am completely calm."

• Refocus attention on your arms. Silently or quietly and slowly repeat to yourself six times, "My arms are very warm." Then silently or quietly say to yourself, "I am completely calm."

• Focus attention on your legs. Silently and quietly and slowly repeat to yourself six times, "My legs are very heavy." Then silently or quietly say to yourself, "I am completely calm."

• Refocus attention on your legs. Silently or quietly and slowly repeat to yourself six times, "My legs are very warm." Then silently or quietly say to yourself, "I am completely calm."

- Silently or quietly and slowly repeat to yourself six times, "My heartbeat is calm and regular." Then silently or quietly say to yourself, "I am completely calm."

- Silently or quietly and slowly repeat to yourself six times, "My breathing is calm and regular." Then silently or quietly say to yourself, "I am completely calm."

- Silently or quietly and slowly repeat to yourself six times, "My abdomen is warm." Then silently or quietly say to yourself, "I am completely calm."

- Silently or quietly and slowly repeat to yourself six times, "My forehead is pleasantly cool." Then silently or quietly say to yourself, "I am completely calm."

- Enjoy the feeling of relaxation, warmth, and heaviness.

- When you are ready, silently or quietly say to yourself, "Arms firm, breathe deeply, eyes open."

In addition to following these instructions, you may consider using a voice recording, such as the free MP3 audio file offered by McMaster University that provides helpful directions for your practice of this technique. Following this audio recording will assist you in fully relaxing so you can concentrate on the technique.

In this chapter I have provided a small sample of key studies from a large and growing body of research. The import of these findings is that our conscious and unconscious beliefs, intentions, attitudes, and emotions—plus any traumatic experiences in our lives or even in the lives of our ancestors—can have a direct, causal, and enduring impact on the DNA of every cell in our bodies. Once a pathway is established in our mental or emotional life through habitual behaviors, it is self-perpetuating and continues its positive or negative influence on our physical and mental health until we intervene to change it. If your aim is to achieve optimal health, it is my firm belief that you must intervene on the side of creating positive changes to your present state of consciousness, which we now know can have long-lasting biological effects.

6

THE ERA OF PERSONALIZED MEDICINE

What the Future Holds

The deterministic model of genetics reigned in the last century, but we've learned in this book how it began to unravel because of the findings of the Human Genome Project. Until the dawn of the epigenetics revolution in the succeeding years, many clinicians thought that medicine was here primarily to save us from bad genes. Most of mainstream medicine was convinced that little could be done about defects in our inherited genome other than some sort of a heroic intervention.

That mentality remains strong even today, and the sad case of Angelina Jolie is illustrative. We learned in Chapter 1 that she had been unfortunate enough to inherit the BRCA mutation. You'll recall that she had her breasts removed because she had been told that she had an 87 percent chance of developing breast cancer, so she and her doctors went ahead with a drastic intervention. But in light of what I have argued in this book, you are now in a position to consider an additional scientific fact: Before 1940, the incidence of breast cancer in such women was only 24 percent.[56] What had changed to elevate their chances more than threefold? Certainly not the gene itself. The important variable, we now know, are the factors influencing this gene's expression: diet, exercise, exposures to pollutants, and other lifestyle behaviors. Therefore, even in Jolie's case—with the odds far more against her in 2013—there were better alternatives than the "heroic" preemptive surgery she elected to undergo.

In the chapters that have followed our examination of her case, we have discovered what makes all the difference. We learned that the deterministic terrain of twentieth-century genetics has now been replaced by the malleable landscape of today's epigenetics. This revelation has in turn brought into being a novel approach to medicine in which the most fundamental tenet is the *person*—that is, the peculiar factor of individuals' highly personal influences on their unique genome.

From all the evidence I have presented, it is clear that our personal beliefs and lifestyle choices create our health reality. There's no way around it; you and I are going to have very different responses to the same inherited genes—we saw this

56 A. Ross, MD, "Genes and Environment in BRCA-Positive Breast Cancer," *ARC Journal of Cancer Science* (2015): 1(1), 17–19, www.arcjournals.org.

vividly in the case of studies of identical twins. Indeed, we are going to have unique responses to *every* sort of life experience. For example, if a group of people hear a sudden loud bang, each of them will respond differently based on their personal history. A returning veteran suffering from PTSD may overreact with fear and dread. A schoolteacher might have pleasant memories of a July Fourth celebration. A person with diminished hearing may not respond at all. Each of us has inimitable reactions that are distinct from the external events around us.

For medicine this means that there is no mythical "average" human. When it comes to health optimization and the routine practice of medical diagnosis and treatment, it is no longer true that "one size fits all." Our challenge today is to jump-start a transformation of healthcare from a medicine designed for a "standard" human to one that treats each individual as special. According to the functional medicine pioneer Dr. Jeffrey Bland,

> Disease is a delusion, one that has been shattered by the still-emerging science of genomics. Breakthrough discoveries over the last decade of the twentieth century and the first decade-plus of the twenty-first have demonstrated that your heart disease is not the same as mine, that everyone with type 2 diabetes is not just like everyone else with type 2 diabetes, that the people with rheumatoid arthritis or Alzheimer's disease are not all similar to others with the same diagnosis. Rather, these

> so-called diseases are dysfunctions of each individual's physiological functioning; they are due to varied causes, and they demand treatment approaches as different from one another as are the individuals.[57]

How we personally respond to our world determines what we uniquely experience in our bodies; this is the epigenetic basis of our health. We are literally creating or undermining our health and longevity with our singular daily experience of life itself.

Our goal, therefore, should be to seize this new understanding and accelerate biomedical progress to create the era of personalized medicine. The next step must be to complete what genetics pioneer Eric Lander calls the "biomedicine's periodic table"—the total reference map of the complete human genome and epigenome with its myriad of functions. According to Lander, this endeavor will take at least another decade of systematic research to define all the elements.[58] The 1000 Genomes Project launched in January 2008 was an international research effort to establish such a map. But the NIH plans to launch an even bigger study known as the All of Us Research Program, which involves a cohort of at least one million volunteers from around the United States. All of

---

57 J. S. Bland, *The Disease Delusion: Conquering the Causes of Chronic Illness for a Healthier, Longer, and Happier Life* (HarperCollins, 2014).

58 Eric S. Lander, "Brave New Genome," *New England Journal of Medicine* (2015), 373:5–8, http://www.nejm.org/doi/full/10.1056/NEJMp1506446?rss=searchAndBrowse&page=&sort=oldest.

this data, once it is processed fully, would set the parameters for highly personalized health practices, medical treatments, and measurement systems that can be adjusted to individual needs almost on the fly.

This chapter will summarize the new possibilities and current challenges arising from such research and point the way to a more hopeful future. Now is the time to lay a foundation for personalized and precision healthcare for the entire planet.

The Role of Technology in the Future of Medicine

Advanced computing has been and will be an integral part of this transformation of medicine. According to noted author and inventor Ray Kurzweil, director of engineering at Google, the healthcare sector has become an integral part of the big data computer revolution. This means that the tools of healthcare will be subject to the exponential rate of change that we observe with all forms of digital technology—or what Kurzweil calls the "law of accelerating returns." This law has already led to gene-sequencing tools that are more than a thousand times more powerful today than those used in the early days of the Human Genome Project. In several years—says Kurzweil in his writings—our research tools will again be another thousand times more powerful, but in two decades it will be a *million times* more powerful.[59]

---

59 See R. Kurzweil, *The Singularity Is Near: When Humans Transcend Biology* (Penguin Books, 2005).

Based on this perspective, let's try to imagine what healthcare could look like in the near future. Precise high-tech genomic assays will provide highly detailed personalized profiles to guide our health choices. But related types of assays will also be crucial. Genomic data will be combined with microbiome assessments that analyze trillions of gut bacteria for markers of epigenetic influence. Complete blood tests will track down hundreds of biomarkers, rather than the usual 20–30 markers, allowing us to know precisely what our bloodstream is carrying to and from our cells.

We should envision that futuristic software running on very high-speed computing platforms will comb through massive combined data sets of gene, blood, gut assays, and other indicators. Clinicians of the future will engage in the fine art of interpreting the patterns that emerge from this mass of biomarker data, or in specific subsets of markers. For example, one blood test under development by Illumina, the world's largest maker of DNA sequencing machines, will be able to detect literally all forms of cancer by searching in the blood for infinitesimal amounts of DNA from cancerous cells. Other providers are working on noninvasive skin patches that will draw microscopic amounts of blood to give us up-to-the-minute information on how we are responding to our medications or health practices. Next-generation digital watches or handheld devices will soon be equipped with sensors and graphic displays to help us understand such health data in real time. Or, nanotechnology sensors that we ingest in sugar pills will transmit health data to our devices.

Some observers point to what has been called the "omics" revolution as a way to characterize the rise of personalized medicine supported by such advanced technology. *Genomics* refers to the improved research, diagnostic testing, and treatments arising from DNA-directed high-tech strategies—which some like to call "precision medicine." *Epigenomics*, as we have seen, is the study of how chemical tags on the human genome can alter gene expression now and in subsequent generations. Another new area of "omics" is *pharmacogenomics*, which identifies gene variants associated with drug metabolism. A burgeoning field that we covered earlier, *nutrigenomics,* has been defined as the influence of nutrients on gene function and health. And finally *microbiomics*, once considered an afterthought, is now front and center in the quest to better understand disease. "There is an enlightened group of early adopters in the biomedical area," says Jeffrey Bland, "who understand that the omics revolution is here to stay and it is more than just a bunch of lab tests. It is really a shifting paradigm about how we see health and disease."[60]

We have a long way to go before such "omic" profiles are perfected, but the future direction of individualized, genetics-based diagnosis and treatment is irreversible. Within the next few years, there will be an unprecedented level of immediate, digital, interactive, detailed, and

---

60 Jeffrey Bland, PhD, and Patrick Hanaway, MD: "Taking the Omics Revolution to the Street," *Integrative Medicine* (Feb. 2015) 14(1): 20–23, https://www.ncbi.nlm.nih.gov/pmc/articles/PMC4566451/.

individualized information made available to clinicians and patients.

Hopefully, we won't become hypochondriacs focused on obsessive testing and anxious over what constitutes normal variations in such biological markers. Instead, my prediction is that we will embrace this information for what it is—a historic opportunity for truly personalized care based on data that offers a reliable guide to optimal health, not just for the "average" individual, but for each unique person.

Resistance to Change from the Old Healthcare Model

A crucial problem we must face, however, is that the promise of such advanced biotechnologies may too often be harnessed to bolster the current dysfunctional model of "disease-centric" medicine.

Consider the serious implications of this trend. I've noted many times that conventional genetic-mapping can only predict a tiny percentage of diseases. Yet, most of the popular gene-testing companies are married to the old medical model of disease diagnosis and treatment. We saw earlier that the size of this new market is expected to reach about $20 billion in a few years and that most of these billions will be spent on predicting the risk of major diseases such as heart disease or cancer. And this new industry is just one small part of the $700 billion biomedical and biotechnology sector that largely serves the agenda of mainstream medicine. For the foreseeable future, most of the investments in this sector will focus on predicting diseases and matching drugs to existing disease conditions. After all, disease-care yields the greatest profit margins, given that our current policies support the logic of the old model. How much better for the world if a primary focus of this giant industry were on *preventive* medicine—that is, creating optimal health by mapping how anticipatory approaches can improve gene expression. Rather than fixing us when we are broken, how much better would it be to unlock those genetic characteristics in individuals that will lead to optimal health!

The harsh truth is that less than three percent of today's healthcare dollar goes to preventive medicine. But even if our regressive healthcare policies were somehow overhauled by fiat, the momentum of the system would continue for a time to go in the wrong direction. The diffusion of personalized and preventive medicine on a large scale will require a major shift in the healthcare infrastructure, along with a huge alteration in the thinking and culture of the medical establishment. This degree of transformation is bound to require years, if not decades, to occur. But don't despair; later in this chapter you will learn how to get access to biomedical services based on the new paradigm.

Americans spend far more per capita on healthcare than every other developed country, but receive worse outcomes. As we continue along this unsustainable path, we can expect that the exciting discoveries of genomics will remain focused on providing new drugs based on genetic assays. On the positive side of this trend, better genetic profiling and screening technologies will allow drug developers to move away from single blockbuster drugs to "minibuster" drugs targeted to more specific conditions.

But yet another problem lurks behind this overemphasis on pharmaceuticals: Even if such a drug-oriented approach was desirable, can the average person even afford such advanced drugs? It is well-known that Americans often pay twice as much as patients in other advanced countries for the same drugs. Plus, a large portion of our population is uninsured or underinsured.

It is also most unfortunate that the healthcare approach I recommend will not have insurance coverage in the near future. Nevertheless, for the educated minority who can afford it, personalized healthcare will open the door to a future focused on prevention of disease and promotion of lifelong optimal health. At its heart will be empowered individuals who can access their digitized health data on demand and who are engaged in discovering the lifestyle approaches best suited to optimizing this data set.

One can envision that today's millennials, as they age, will increasingly be interested in assembling personalized healthcare databases, including individual genetic profiles, biomic assays, epigenetic maps, family histories, current and past treatment protocols, and personal preferences. They will create health portfolios analogous to financial portfolios and manage them with the help of wellness managers, health coaches, and integrative medicine clinicians. And at least one aspect of this trend is now going mainstream: The United States is investing billions today in implementing electronic medical records (EMR) as a permanent feature of healthcare.

At some point, medicine *will* shift from its reactive and disease-based focus. Healthcare expenditures *will* move away from the costly and fragmented treatment of disease events that are largely preventable. Personalized approaches with intense patient engagement *will* one day be proven more effective than the current episodic treatment of disease occurrences. We are long overdue for such a seismic shift in our approach to medicine.

The Genome Editing Revolution

At the same time, let's not forget that "integrative care" harvests the best of both worlds of medicine. Integrative medicine is obliged to turn to mainstream allopathic approaches when preventive care fails to stop the onset of chronic illnesses through negligence or misapplication, or when genetic defects are fully penetrant. When lifestyle medicine falls short, most of us will gratefully resort to one or another high-tech medical intervention. In a few more years, one of the more promising of these "heroic" interventions will involve *gene editing.*

To understand the genome-editing revolution, let's step back a bit. Fifty years ago microbiologists first learned that bacteria carry enzymes that, in effect, can be used to "cut and paste" DNA. This discovery enabled *recombinant DNA* to be created in vitro in laboratories and led to the first wave of genetic engineering. This advance in turn transformed genetics, eventually giving rise to today's global biotech industry. Beginning less than a decade ago, another round of discoveries, also related to the enzyme action of bacterial DNA, caused yet another revolution. We now know that bacteria contain a *programmable* mechanism that can direct enzymes to "cut" or alter DNA in surgically precise regions of the genome, like a molecular scissors. The now widely used acronym for this technology, CRISPR, refers to a general-purpose tool for editing the genome in living human cells that already has a myriad of applications; indeed, thousands of laboratories around the world now use CRISPR in their research. One lab at the University of Michigan recently discovered by accident that CRISPR can even edit RNA sequences.[61]

By all accounts, gene editing holds monumental therapeutic promise in those cases that merit a radical intervention. Here are a few illustrative examples:

- The CCR5 gene confers resistance to HIV, but one percent of the U.S. population lacks this gene. Scientists are working on ways to edit a patient's immune cells to insert the missing gene.

61 "New CRISPR-Cas9 Tool Edits Both RNA and DNA Precisely," *Science Bulletin* (Feb. 16, 2018), https://sciencebulletin.org/archives/20408.html.

• A certain mutation found in some unfortunate people prevents their bodies from easily removing LDL cholesterol (the "bad" cholesterol) from the blood. Geneticists believe they will soon be able to edit liver cells to restore normal cholesterol processing to these people.

• Other researchers are working on editing blood stem cells in order to cure sickle cell anemia and hemophilia. A related group is testing a CRISPR treatment that enhances cancer-fighting genes in the blood that can ward off blood cancers.

• A certain inherited gene is known to always produce muscular dystrophy. Researchers have demonstrated how to turn off that gene in mice with the use of CRISPR tools.

• Malaria researchers are looking at a myriad of ways for using CRISPR to introduce genetic interventions in mosquito populations that will reduce the spread of the disease, including sterilizing female mosquitoes so they won't reproduce.

CRISPR editing is known to be very precise, even at this early stage in the laboratory research. But great care must obviously be taken when these tools are moved out of the lab and into clinical practice.

The research projects listed above do not pose difficult ethical challenges because these changes affect only the individual patient who is treated. But what if CRISPR is turned to creating "designer babies" or "genetically modified humans"? Such children—if they survive to become reproducing adults—would carry heritable changes in their DNA, which is a scary prospect. In 2015, several prominent groups in the emerging gene-editing field urged a global moratorium on this form of human DNA editing. In the U.S., the NIH issued strict guidelines governing this sort of research, and the FDA stated that it will not permit genetic modification of embryos. Some European countries have outlawed genetic modification of human embryos.

But in many other countries the situation remains unresolved. For example, an audacious effort to modify human embryos using CRISPR occurred in China in 2015 at Sun Yat-sen University in Guangdong. These scientists attempted to edit 85 defective embryos, but their experiment signally failed. Either the targeted genes were not altered or the edited embryos themselves died.

In an important 2015 editorial published in the *New England Journal of Medicine*, Dr. Eric Lander, director of the Broad Institute at Harvard and MIT, offers a practical starting point for ethical reflection about embryo editing.

> Today, the most common argument for DNA editing concerns preventing devastating monogenic diseases, such as Huntington's disease. Though avoiding

> the roughly 3600 rare monogenic disorders caused by known disease genes is a compelling goal, the rationale for embryo editing largely evaporates under careful scrutiny. . . . To reduce the incidence of monogenic disease, what's needed most is not embryo editing, but routine, preventive genetic testing [of] the many couples who don't know they are at risk.[62]

If we set aside the foolish notion of editing human embryos or even the more useful and feasible idea of targeting rare single-gene disorders, what about using gene-editing tools to reduce the risk of common diseases like heart disease or cancer? This option turns out to be far less viable than one might think, and for a relatively simple reason. We've noted earlier that most disease risk is shaped by variants in *numerous* genes, sometimes numbering in the hundreds. According to Lander, "such variants tend to make only modest contributions [to a given disease]. . . . The epigenetic effects of diet, stress, the environment, and other factors interacting with the genes are [the] major determinants."

And with this statement Lander brings us right back to the core argument of this book: What we do, think, and how we live is the key to working with our inherited genome.

---

62 Eric S. Lander, PhD, "Brave New Genome" (July 2, 2015), http://www.nejm.org/doi/full/10.1056/NEJMp1506446.

Promising Companies of the Future

Twenty-five years ago, biologists weren't even sure of the value of sequencing the human genome. But consider what things must be like for today's rising young geneticists. I would image that they regard these early days as some sort of antideluvian era, before today's flood of genomics data. Genomics is changing the practice of healthcare in fundamental ways, and the most promising result—at least for our purposes in this book—are new companies that are pointing the way forward to new methods of health optimization based on the epigenetic revolution.

Prominent among these pioneers is a San Francisco start-up company called WellnessFX (WFX), founded by Jim Kean. Prior to starting WFX, Kean had been a venture capitalist and a well-respected entrepreneur whose first company was bought out by WebMD. WellnessFX was born in 2008 out of Kean's personal frustration with a bad cholesterol score from a routine exam. When his doctor offered statins, Kean hesitated. He was an avid athlete, and he also knew that statins had considerable side effects and might slowly degrade his muscle fibers. What other biomarkers, he wondered, could give him insight into his condition? After getting his doctor to run a bigger panel of tests, Kean discovered that *one-third* of his biomarkers were in the risk range.

He dove into research, motivated by his new health concerns. Kean soon realized that many middle-aged people like himself are disengaged from tracking their health issues because the results of standard blood tests are hard to interpret and are too often focused on "disease-care" rather than prevention. He founded WellnessFX to design a platform that addressed those problems through advanced blood tests that provide a path to advanced personalized care. In 2015, after Kean's company was purchased by a large and innovative nutraceutical company called Thorne Research, something unique began to unfold.

Thorne and its affiliates are in my view "best-in-class" in this emerging field of biomarker testing for health optimization. By way of disclosure, I am one of Thorne's compensated advisers, but my remuneration is not tied to Thorne's market success. Thorne is best known as the manufacturer of high-grade nutritional supplements that it sells to doctors and elite athletes. It has partnered with the Mayo Clinic on numerous research projects and is the only supplement company that is permitted to display the Olympic logo. Most recently, Thorne purchased a company called Pillar Health, with the goal of building an artificial intelligence platform that analyzes the results of biomarker testing.

WFX, through its partnership with Thorne, has become an industry leader in offering an advanced tripartite assay focused on prevention. This package will use artificial intelligence to merge the data of three very different categories of biomarkers into a coherent profile that is delivered to the customer along with actionable health recommendations. The three components are:

1.) Genomics—A gene profile is created that is focused on function and fitness markers (rather than disease markers), all of which are modifiable through lifestyle changes. Pathway Genomics, the genetic testing contractor to WFX, focuses on 75 genetic markers known to affect metabolism, exercise, and energy use, and this test provides a detailed analysis of how an individual's body responds to exercise.

2.) Blood or CBC—This entails a complete blood chemistry (CBC) test that covers approximately 75 biomarkers. Originally developed by WFX, the test measures lipids (fats) and lipid subfractions, homocysteine, steroid hormones (such as testosterone), iron, B12, liver function, and kidney function.

3.) Biomic or Gastrointestinal—This test of the gut biome is based on a detailed stool analysis that focuses on hundreds of biomarkers related to metabolism and digestion. These specialized assays are conducted by the Mayo Clinic and Thorne's partner, the laboratory of Dr. Chris Mason at the Weill-Cornell Medical Center (www.masonlab.net).

Thorne/WFX sends kits to the home for all three tests and provides consumers with an online dashboard displaying the data in a form that is meaningful to the layperson. According to the president of Thorne, Paul Jacobson, the aim of the company and its partners is to provide four deliverables: an advanced tripartite test; an analysis of the results through artificial intelligence; a simple and accessible display of its results; and a formal set of preventive suggestions that include food, supplements, and exercise. "We offer a holistic approach where you get everything done in one place," says Jacobson. "Our tripartite assay reveals altered or missing molecular components in the body that must be addressed in order to accelerate wellness. If any of the biomarkers point to a disease threat, we immediately refer our clients to their physician."

Another such company, Habit, was founded by Neil Grimmer, a vegan Ironman triathlete. Neil knew the power that food had to unlock his full potential in life and on the race course. When he became a father, he started Plum Organics, which today is a leading baby food brand across America.

In nurturing Plum, Neil experienced what a lot of new parents go through: His own well-being took a hit. He found himself 65 pounds heavier than when he was racing, with health issues that were challenging his longevity. After DNA and blood tests, he discovered that the best way back to a healthier, energized version of himself started with understanding his body's fundamental needs and eating the types of foods and nutrients that his body uniquely asked for. Twenty-five pounds lighter, and feeling better than ever, Neil set out to build a nutrition company based on making optimal health and well-being possible through holistic, personalized nutrition guidance.

This important new company uses three blood samples

and a DNA swab, all collected at home, to generate advice based on a "systems biology methodology." First, the inside of the cheeks is thoroughly swabbed, after which the three simple blood samples (by pricking fingertips) are drawn. These draws are interspersed over several hours; they are taken before—and at 30-minute and two-hour intervals after—the consumption of a special shake, which is "clinically validated to help assess changes in nutrition-related blood biomarkers." These samples are mailed in along with a detailed questionnaire, and the test data are run through a "nutritional analysis algorithm" that analyzes 60-plus indicators that include gene variations, blood markers, metabolism, and body metrics related to nutrition and general health. Clients learn things such as their genetic likelihood to have lactose intolerance or caffeine sensitivity, or whether they have a propensity to weight gain. Working with a coach, customers develop a nutrition plan that includes determining the best foods for their genetic type and the ideal balance of protein, carbohydrates, and fats.

A number of other companies, many with great scientists, have also entered the biomarker testing space. These can be broken down into (1) those focused on disease-state analytics, and (2) those interested in promoting prevention and wellness via advanced testing platforms. Some companies combine the two. Many teams are looking at creating some form of tripartite test for the consumer, while some focus on just one assay (such as biomic) or two assay types (e.g., a blood and a biomic test). The field is changing fast, but companies currently of note include the following:

23andMe: This well-known company, having reformed itself after being shut down by the FDA (see Chapter 2), is now teamed with Ancestry.com to offer a "health + ancestry" package for under $200 that is based on a simple home-based saliva collection kit. Their ancestry reports cover five aspects of "ancestry composition," and they provide 40+ "carrier status" reports related to ancestry that indicate whether the client carries well-known disease variants such as sickle cell anemia or cystic fibrosis. Their small set of "health risk reports" include macular degeneration, Celiac disease, and Alzheimer's disease. Their "wellness reports" cover alcohol, caffeine, and lactose tolerance, and a handful of other issues. 23andMe is the only player in this space that combines ancestry genetics with a genetic health profile.

Human Longevity: Founded by Dr. J. Craig Venter, a prominent pioneer in human genome sequencing, this company combines whole-genome sequencing with a broad set of other clinical and biological measures including microbiome testing; a process termed "metabolome characterization"; and screenings that include a whole-body MRI. It entails a high price tag, but their approach is advanced and comprehensive.

Viome: Formed by a well-known scientific team, this company tests the genetic expression of the gut microbiome using a stool collection kit and combines this with a test performed at home to determine how the customer metabolizes key macronutrients. The company sorts these results to determine the client's "ideal macronutrient ratio," which results in customized dietary recommendations. "By analyzing the genes that your gut microbes express, we can identify which metabolites they produce," states Viome. "By following Viome's diet and lifestyle recommendations, you'll be able to fine-tune the function of your gut microbiome to minimize production of harmful metabolites and maximize the production of beneficial ones."

uBiome: This well-regarded company offers a microbiome test using a "tissue swipe" rather than a stool sample. Among its offerings, the company claims to provide "the world's first gene-sequencing-based clinical microbiome test" that is primarily aimed at helping the client and their doctor manage gut conditions like inflammatory bowel disease (IBD) and irritable bowel syndrome (IBS).

Arivale: Founded by a strong scientific team, for a fee of $999 this company tests 80 biomarkers with a blood test, and offers a proactive coach to help you interpret results and pursue lifestyle changes.

Let me be very clear that there are many, many caveats in direct-to-consumer genetic testing. It is now 65 years since James Watson, Francis Crick, Maurice Wilkins, and Rosalind Franklin determined the double helix structure of DNA. One rightfully skeptical individual is Lisa Suennen, who leads the GE Ventures healthcare venture capital program and has over 30 years of venture capital experience. In her blog, she noted, "Obviously there are massive differences between tumor sequencing when someone has advanced cancer and consumer-directed genetic tests. We should be grateful that the former is becoming commonplace, as it will likely create massive benefit. As for the latter, well . . . it remains to be seen, except for the entertainment value; or finding a roommate." She also astutely observes, "But that [database volume] assumes that companies like 23andMe and others will sell their data to pharmaceutical companies (oh wait, they do that) and that consumers will be ok with that."[63] As we have seen recently with the Facebook data breach scandal involving over 87 million Facebook users, unauthorized, nondisclosed use of private data has raised major ethical, legal, and moral questions that will take many years to address. With this major caveat clearly in mind, it is my opinion that there are companies offering and will be offering genetic assays that will in fact improve individual health status.

Back to the Future

This section could be titled "from outer space to inner space" and back again! Just as I was finishing this book, there was a press release from NASA that was such a perfect instance of epigenesis at work that I laughed out loud. It focused on the return

63 Lisa Suennen, "Genomics Rising, Charlatans Circling—Marketing Claims Growing out of Control," *Medpage Today* (Apr. 4, 2018).

of Astronaut Scott Kelly, who set a record for the longest solo spaceflight in history. Scott has a twin brother named Mark, who is also a NASA astronaut but remained on Earth during his twin brother's stay aboard the international space station. NASA compared their DNA after Scott spent nearly a year (340 straight days) in space. Initially, the study indicated that Scott had grown over 2 inches in height during zero gravity.

Even more surprising was that Scott's telomeres actually became significantly longer in space. NASA also revealed that Scott had hundreds of "space genes," which were changed in their expression by the yearlong space flight. These epigenetic changes included gene expressions affecting the "immune system, DNA repair, bone formation networks, hypoxia, and hypercapnia." After his return to Earth, his height changed back to normal, as did 93 percent of his DNA expression. However, the epigenetic expression of 7 percent of his genes remained changed and might remain that way permanently. According to the press release, "This is thought to be from the stress of space travel, which can cause changes in a cell's biological pathways and ejection of DNA and RNA."[64] In other words, these potentially permanent changes in gene expression are perfect demonstrations of the principles of epigenetics.

In a posting on Twitter, Scott quipped, "What? My DNA changed by 7%! Who knew? I just learned about it in this article. This could be good news! I no longer have to call @ ShuttleCDRKelly my identical twin brother anymore." He is

64 CBS Denver, "Astronaut Scott Kelly Now Has Different DNA Than His Identical Twin Brother after One Year in Space" (Mar. 13, 2018).

partly right in that his DNA did not actually change, but the expression of 7 percent of his DNA did change in a manner consistent with epigenesis. Clearly, as we venture to the moon again and later to Mars and beyond, we will literally be moving from outer space to inner space, as deep space travel exerts a profound influence on our core DNA and its expression.

Back here on Earth, two fundamental ideas underlie the brave new era of personalized health care. First, we are able to identify with great precision strengths and weaknesses that are written into each person's inherited DNA. And second, we can determine the specific choices we can make in diet, stress management, environment, and other lifestyle factors that will express or suppress those genetic tendencies. In clinical practice, this means we can couple established genetic indicators with state-of-the-art individual profiles to create diagnostic, prognostic, and therapeutic strategies precisely tailored to each individual's requirements. And this new regime of precision health care will span the entire spectrum from wellness and optimal health to effective disease treatment.

Those of us who aspire to be active participants in optimizing our health should be aware that a profound revolution in true healthcare based on genomics is now upon us. It is up to all of us to ensure that this powerful knowledge lives up to its promise. We truly stand at the brink of an age of optimal health and longevity that is unparalleled in human history. At the heart of the personalized healthcare system that will support this new epoch stands the individual who is engaged in his or her own well-being in a way that is cost-effective, sustainable, and will be a boon to our personal as well as national health and economy.

Given the thrilling advances covered in this book, it seems possible to at last bridge the gap between our spiritual purpose and the age-long quest for human health and longevity. But the brilliant advances in biomedical technology alone cannot provide all the answers we are seeking. Twenty-first-century science has permitted humankind to peer with electron microscopes into intercellular space and chart the helical coils of the DNA molecules at the very heart of all life. Yet, ancient scriptures long ago charted the journey toward wisdom and enlightenment in the mind and inward to the very heart of the human soul. Whether we are scientists or mystics, the enduring mysteries of life on this Earth should instill in all of us an abiding sense of awe, humility, and compassion.

A SELECTION OF SOURCES CONSULTED

Chapter 1

Devey, F. E., et al. "Clinical Interpretation and Implications of Whole Genome Sequencing. *JAMA* 311 (March 12, 2014): 1035.

Ecker, J. R., et al. "Genomics: ENCODE Explained." *Nature* 489 (September 6, 2012): 52.

FDA Warning Letter to "Twenty Three and Me." November 22, 2013, 1–4. http://www.fda.gov/ICECI/EnforcementActions/WarningLetters/20`3/ucm376296.htm.

Ferro, W. G. "Clinical Applications of Whole Genome Sequencing: Proceed with Care." *JAMA* 311 (March 12, 2014): 1047.

Ford, Earl S., Manuela M. Bergmann et al. "Healthy Living Is the Best Revenge—Findings from the European Prospective Investigation into Cancer and Nutrition–Potsdam Study." *Archives of Internal Medicine* 169, no. 15 (2009): 1355–1362. doi:10.1001/archinternmed.2009.23.

Gutierrez, Alberto. Warning Letter. 23andMe, Inc., November 22, 2013. Inspections, Compliance, Enforcement and Criminal Investigations. US Food and Drug Administration. http://www.fda.gov/ICECI/EnforcementActions/WarningLetters/2013/ucm376296.htm.

Manolio, T. A. "Genomewide Association Studies and Assessment of the Risk of Disease." *New England Journal of Medicine* 363, no. 12 (July 8, 2010): 166–176.

Pennisi, E. "ENCODE: Project Writes Eulogy for Junk DNA." *Science* 337 (September 7, 2012): 1159.

Rabin, Roni Caryn. "In Israel, a Push to Screen for Cancer Gene Leaves Many Conflicted." *New York Times*, November 26, 2013. http://www.nytimes.com/2013/11/27/health/in-israel-a-push-to-screen-for-cancer-gene-leaves-many-conflicted.html?_r=0.

Roy-Byrne, P. "Emerging Perspectives: Update on Genome-Wide Association Studies." *Journal Watch Online*, April 20, 2009. http://psychiatry.jwatch.org/cgi/content/full/2009/417/4?q=etoc_jwpsych.

Saey, T. "People Have a Surprising Number of Rare Genetic Variants." *Science News*, August 25, 2012, 28–29. www.sciencenews.org.

Scheuner, M. T., P. Sieverding, and P. G. Shekelle. "Delivery of Genomic Medicine for Common Chronic Adult Diseases: A Systematic Review." *JAMA* 299, no. 11 (March 19, 2008): 1320–1334.

Chapter 2

Cai, N., et al. "Molecular Signatures of Major Depression." *Current Biology* 25, no. 9 (May 4, 2015): 1146. dx.doi.org/10.1016/j.cub.2015.03.008.

Herman, A. O. "US Moves to Ensure Accuracy of Genetic Test Results." *Journal Watch*, May 18, 2015. http://jwatch.org/fw110208/2015/05/18/us-moves-ensure-accuracy-genetic-test-results?query=pfwTOC#sthash.vU7nrl6Z.dpuf.

Kolata, G. "Bits of Mystery DNA, Far from 'Junk,' Play Crucial Role." *New York Times*, September 5, 2012.

Kolata, G. "Chinese Scientists Edit Genes of Human Embryos, Raising Concerns." *New York Times Online*, November 23, 2015.

Kurzweil, R. "This Is Your Future." Special to CNN Online, December 10, 2013.

Rampazzo, E., et al. "Relationship between Telomere Shortening, Genetic Instability, and Site of Tumor Origin in Colorectal Cancers." *British Journal of Cancer* 102, no. 8 (April 13, 2010): 1300–1305.

Riddihough, G. "Spreading Small RNAs to Protect the Genome." *Science* 348, no. 6236 (May 15, 2015): 768.

Szyf, M. "Lamarck Revisited: Epigenetic Inheritance of Ancestral Odor Fear Conditioning." *Nature Neuroscience* 17, no. 2 (January 2014). dx.doi.org/10.1038/nn.3603.

Chapter 3

Herman, A. O. "US Moves to Ensure Accuracy of Genetic Test Results." May 18, 2015. http://jwatch.org/fw110208/2015/05/18/us-moves-ensure-accuracy-genetic-test-results?query=pfwTOC#st-hash.vU7nrl6Z.dpuf.

Jagsi, R., et al. "Concerns about Breast Cancer Risk and Experiences with Genetic Testing in a Diverse Population of Patients with Breast Cancer." *Journal of Clinical Oncology* 33, no. 14 (May 10, 2015): 1584. dx.doi.org/10.1200/JCO.2014.58.5885.

McGrath, M., et al. "Telomere Length, Cigarette Smoking, and Bladder Cancer Risk in Men and Women." *Cancer Epidemiology, Biomarkers and Prevention* 16, no. 4 (April 2007): 815–819.

Milagro, F. I., et al. "A Dual Epigenomic Approach for the Search of Obesity Biomarkers: DNA Methylation in Relation to Diet Induced Weight Loss." *FASEB Journal* 25, no. 4 (April 2011): 1379–1389. doi:10.1096/fj.10-170365.

Riddihough, G. "Spreading Small RNAs to Protect the Genome." *Science* 348, no. 6236 (May 15, 2015): 768.

Terry, D. F., et al. "Association of Longer Telomeres with Better Health in Centenarians." *Journals of Gerontology, Series A: Biological Sciences and Medical Sciences* 63, no. 8 (August 2008): 809–812.

Chapter 4

Action to Control Cardiovascular Risk in Diabetes Study Group, Group, H. C. Gerstein et al. "Effects of Intensive Glucose Lowering in Type 2 Diabetes." *New England Journal of Medicine* 358, no. 24 (June 12, 2008): 2545–2559.

Chandalia, M., A. Garg, and D. Lutjohann. "Beneficial Effects of High Dietary Fiber Intake in Patients with Type 2 Diabetes Mellitus." *New England Journal of Medicine* 342, no. 19 (May 11, 2000): 1392–1398.

Chen, L., L. J. Appel et al. "Reduction in Consumption of Sugar-Sweetened Beverages Is Associated with Weight Loss: The PREMIER Trial." *American Journal of Clinical Nutrition* 89, no. 5 (May 2009): 1299–306.

Cordain, L., et al. "Origin and Evolution of the Western Diet: Health Implications for the Twenty-First Century." *American Journal of Clinical Nutrition* 8, no. 2: 341–354.

Estruch, Ramón, Emilio Ros et al. for the PREDIMED Study Investigators. "Primary Prevention of Cardiovascular Disease with a Mediterranean Diet." *New England Journal of Medicine* 368 (April 4, 2013): 1279–1290. doi:10.1056/NEJMoa1200303.

Ford, Earl S., Manuela M. Bergmann et al. "Healthy Living Is the Best Revenge—Findings from the European Prospective Investigation into Cancer and Nutrition–Potsdam Study." *Archives of Internal Medicine* 169, no. 15 (2009): 1355–1362. doi:10.1001/archinternmed.2009.23.

Hu, F. B., T. Y. Li et al. "Television Watching and Other Sedentary Behaviors in Relation to Risk of Obesity and Type 2 Diabetes Mellitus in Women," *JAMA* 289, no. 14 (April 9, 2003): 1785–1791.

Kelly, G. S. "Insulin Resistance: Lifestyle and Nutritional Interventions." *Alternative Medicine Review* 5, no. 2 (April 2000): 109–132.

Kligler, B., and D. Lynch. "An Integrative Approach to the Management of Type 2 Diabetes Mellitus." *Alternative Therapies in Health and Medicine* 9, no. 6 (November–December 2003): 24–32, quiz 33.

Labonté, B., et al. "Genome-Wide Epigenetic Regulation by Early-Life Trauma." *Archives of General Psychiatry* 69 (July 2012): 722. https://www.ncbi.nlm.nih.gov/pmc/articles/PMC4991944/.

Lakka, H. M., D. E. Laaksonen et al. "The Metabolic Syndrome and Total and Cardiovascular Disease Mortality in Middle-Aged Men." *JAMA* 288, no. 21 (December 4, 2002): 2709–2716.

Ludwig, D. S., K. E. Peterson, and S. L. Gortmaker. "Relation between Consumption of Sugar-Sweetened Drinks and Childhood Obesity: A Prospective, Observational Analysis." *Lancet* 357, no. 9255 (February 17, 2001): 505–508.

Robson, A. A. "Preventing Diet Induced Disease: Bioavailable Nutrient-Rich, Low-Energy-Dense Diets." *Nutrition and Health* 20, no. 2 (2009): 135–166.

Ryan, A. S. "Insulin Resistance with Aging: Effects of Diet and Exercise." *Sports Medicine* 30, no. 5 (November 2000): 327–346.

Willeit, P., et al. "Telomere Length and Risk of Incident Cancer and Cancer Mortality." *JAMA* 304 (July 7, 2010): 69. http://dx.doi .org/10.1001/jama.2010.897.

Chapter 5

Amstadter, A. B., et al. "Psychiatric Resilience: Longitudinal Twin Study." *British Journal of Psychiatry* (October 2014): 275. http://dx .doi.org/10.1192/bjp.bp.113.130906. See more at http://www.jwatch .org/na35944/2014/10/23/genetic-contributors-resilience?query=etoc_jwpsych#sthash.4Ma71srv.dpuf.

Atasoy, Ozgun. "Your Thoughts Can Release Abilities beyond Normal Limits." *Scientific American*, December 16, 2013.

Bhasin, M. K., et al. "Relaxation Response Induces Temporal Transcriptome Changes in Energy Metabolism, Insulin Secretion, and Inflammatory Pathways." *PLoS ONE* 8 (May 1, 2013): e62817.

Eley, T. C., et al. "The Intergenerational Transmission of Anxiety: A Children-of-Twins Study." *American Journal of Psychiatry* 172, no. 7 (April 23, 2015): 630–637. http://dx.doi.org/10.1176/ appi.ajp.2015.14070818. See more at http://www.jwatch.org/ na37764/2015/05/18/anxiety-disorders-environmental-or-genetic-transmission?query=etoc_jwpsych#sthash.6tn0VpIq.dpuf.

Eley, T. C., et al. "Therapygenetics: The 5-HTTLPR and Response to Psychological Therapy." *Molecular Psychiatry* (October 25, 2011). http://dx.doi.org/10.1038/mp.2011.132.

El Kordi, A., et al. "A Single Gene Defect Causing Claustrophobia." *Translational Psychiatry* 3 (April 30, 2013): e254. http://dx.doi.org/10.1038/tp.2013.28.

Flint, J., and M. Munafò. "Genesis of a Complex Disease." *Nature* 511 (July 24, 2014): 412. http://dx.doi.org/10.1038/nature13645. See more at http://www.jwatch.org/na35488/2014/08/21/huge-study-elucidates-genetics-schizophrenia?query=etoc_jwgen-med#sthash.8M9Dtz1T.dpuf.

Gollub, Randy L., and Jian Kong. "For Placebo Effects in Medicine, Seeing Is Believing." *Science Translation Medicine* 70 (February 16, 2011): 5.

Gray, R. "Phobias May Be Memories Passed Down in Genes from Ancestors." *Telegraph Sun*, May 24, 2015, 1–3.

Kaliman, Perla, et al. "Rapid Changes in Histone Deacetylases and Inflammatory Gene Expression in Expert Meditators." *Psychoneuroendocrinology* 40 (February 2014): 96–107. dx.doi.org/10.1016/j.psyneuen.2013.11.004.

Labonté, B., et al. "Differential Glucocorticoid Receptor Exon 1B, 1C, and 1H Expression and Methylation in Suicide Completers with a History of Childhood Abuse." *Biological Psychiatry* 72 (July 1, 2012): 42.

Miller, G. E., and S. W. Cole. "Clustering of Depression and Inflammation in Adolescents Previously Exposed to Childhood Adversity." *Biological Psychiatry* 72 (July 1, 2012): 34.

Ornish, D., J. Lin, et al. "Effect of Comprehensive Lifestyle Changes on Telomerase Activity and Telomere Length in Men with Biopsy-Proven Low-Risk Prostate Cancer: 5-Year Follow-up of a Descriptive Pilot Study." *Lancet* 14, no. 11 (September 17, 2013): 1112–1120. dx.doi.org/10.1016/S1470-2045(13)70366-8.

Qiu, A., et al. "Prenatal Maternal Depression Alters Amygdala Functional Connectivity in 6-Month-Old Infants." *Translation Psychiatry* 5 (February 17, 2015): e508. http://dx.doi.org/10.1038/tp.2015.3. See more at http://www.jwatch.org/na37149/2015/03/11/prenatal-depression-affects-brain-functional-connectivity#sthash.YFOOPxCT.dpuf.

Schizophrenia Working Group of the Psychiatric Genomics Consortium. "Biological Insights from 108 Schizophrenia-Associated Genetic Loci." *Nature* 511 (July 24, 2014): 421. http://dx.doi.org/10.1038/nature13595.

Shalev, I., et al. "Exposure to Violence during Childhood Is Associated with Telomere Erosion from 5 to 10 Years of Age: A Longitudinal Study." *Molecular Psychiatry* 18 (May 2013): 576. https://www.nature.com/articles/mp201232.

Stetter, F., and S. Kupper. "Autogenic Training: A Meta-analysis of Clinical Outcome Studies." *Applied Psychophysiology and Biofeedback* 27, no. 1 (2002): 45–98.

Chapter 6

Bland, J. *The Disease Delusion: Conquering the Causes of Illness for a Healthier, Longer, and Happier Life*. Harper Wave, 2014.

Lander, Eric S. "Brave New Genome." *New England Journal of Medicine* 373 (June 3, 2015). doi:10.1056/NEJMp1506446.

INDEX

An "n" followed by a page number indicates a footnote on the bottom of that page.